CAMBRIDGE COUNTY GEOGRAPHIES

General Editor: F. H. H. Guillemard, M.A., M.D.

BRECONSHIRE

Cambridge County Geographies

BRECONSHIRE

by

CHRISTOPHER J. EVANS

With Maps, Diagrams and Illustrations

Cambridge:
at the University Press
1912

CAMBRIDGE UNIVERSITY PRESS
Cambridge, New York, Melbourne, Madrid, Cape Town,
Singapore, São Paulo, Delhi, Mexico City

Cambridge University Press
The Edinburgh Building, Cambridge CB2 8RU, UK

Published in the United States of America by Cambridge University Press, New York

www.cambridge.org
Information on this title: www.cambridge.org/9781107629219

First published 1912
First paperback edition 2013

A catalogue record for this publication is available from the British Library

ISBN 978-1-107-62921-9 Paperback

PREFACE

THE author desires to acknowledge his indebtedness to several works on the history and antiquities of Breconshire, especially to *The Birds of Breconshire* by Mr E. Cambridge Phillips. His thanks are also due to Mr John Ward, F.S.A., for his great assistance during the photographing of exhibits in the Welsh Museum, Cardiff; to Mr C. H. Priestley, M.I.C.E., for plans and information supplied; to Dr W. Black Jones of Llangammarch Wells for kindly furnishing sunshine statistics; to the Principal of the Cardiff University College for permission to photograph the Celtic handbell; and to several ladies and gentlemen of Breconshire who readily gave permission for photographs to be taken when they understood the object for which they were intended.

C. J. EVANS.

January 1912.

SIR FRYCHEINIOG

Wrth ochr hon mae Sir Frycheiniog,
Mynyddoedd mawr a mannau cribog,
Ym Muallt rhosydd tiroedd oerllyd,
Rhai mynyddau a choedydd hefyd.

Tiroedd da a choed ddiogel,
O'r Gelli i Dalgarth Cerrig Hywel,
Ac o gwmpas Aberhonddu,
Dyna'r dref gyfoethoca yn Nghymru

Ac yn hon mae Protestaniaid
A thair eglwys iddynt fyned,
Ond hyd eu gwlad mewn amryw fannau,
Mae rhai'n pregethu hyd eu teiau.

Haidd a gwenith, a hefyd rygau
Caws a 'menyn sydd mewn mannau;
Purion bara ceirch diogel,
Yn Nghwmwd Muallt ac yn Llywel.

Gwna'r merched hyn yn gofus gyfan,
Bob gwaith hyswi i mewn ac allan;
Gwau'r hosanau drwy'r holl flwyddyn,
O Lan-Fair hyd yn Aber-Gwesyn[1].

1814. D. THOMAS A'I CANT.

[1] Contiguous to Radnorshire lies Breconshire, mountainous with many precipitous places. Builth (the commot) has meadows, exposed lands, some mountains and also forests.

Good lands and plenteous forests extend from Gelli to Talgarth and Crickhowell, and also around Brecon :—that is the wealthiest town in Wales.

The Protestants here have three churches, but over the countryside, in several places, some preach from house to house.

Barley and wheat and also rye, cheese and butter, are to be found in some places. Plenty of good oaten bread in the commot of Builth and in Llywel.

The maids of Breconshire attend carefully to duties within and without the house, and through the whole year, in the district between Llanfair and Abergwessyn, they knit stockings.

CONTENTS

ILLUSTRATIONS

MAPS

The illustrations on pp. 4, 13, 19, 26, 45, 103, 118, 128, 136, 152 are from photographs supplied by Messrs F. Frith & Co., Ltd., of Reigate; that on p. 74 by Mr P. B. Abery of Builth; that on p. 91 by Mr R. H. Thomas of Aberdare; that on p. 165 by Mr R. H. Stevens of Crickhowell; and the remainder, with the exception of that on p. 159, from photographs taken expressly for this book by Mr Fred Evans, Llangynwyd, Glamorganshire.

1. Siluria. Its division into Principalities. Garth Madryn, Brycheiniog, Brecknock, and Breconshire.

When the Romans made their appearance in the islands of Britain, the district now forming the south-eastern corner of Wales was known to the Britons as Essyllwg, and was inhabited by a tribe known as Essyllwyr. These names the Romans translated into Latin as Siluria and Silures respectively. Siluria comprised most of the present counties of Glamorgan and Breconshire, the whole of Monmouthshire, and some other outlying tracts now forming parts of English counties. It was divided into districts, probably under sub-chiefs, and of these a tract of country called Garth Madryn was one. We do not know the actual extent of this district, but, roughly, it was contained within the boundaries that now mark the present county of Breconshire, with the possible exception of the hundred of Builth.

Garth Madryn, we are told, means "Fox Hill" or "Fox Hold," and was thus named because it formed the home of large numbers of those animals. Perhaps the name was only given to a small portion of the district at first and was afterwards extended to include the whole area we have outlined.

Still this does not help us to understand how the county received its present name. Neither does it tell us what it means. For this information we must come to a period several hundreds of years later, in fact to the period of the departure of the Romans. About this time the Welsh prince of Garth Madryn was a man named Brychan, and in his honour it was renamed Brycheiniog or Brycheiniawg—that is, the Land of Brychan. From that time onwards its princes or rulers were always described as of Brycheiniog, and in later times still, when the Normans took most of the district from the Welsh, the name was still retained to designate the Lordship formed by the Norman conqueror who was known as the Lord of Brycheiniog.

The Normans half-translated the word Brycheiniog, and after some changes the word they used in time became fixed as Brecknock, or as it is nowadays written, Brecon. To the Welsh the name is still Brycheiniog, and when we speak of Breconshire we say Sir Frycheiniog. The difference in spelling is simply due to a grammatical rule which alters the first letters of many words in the Welsh language. For a long period the names Breconshire and Brecknockshire have been used indiscriminately for our county, but in 1910 the County Council, by resolution, determined that the form Breconshire shall be used in all official documents. This may have the effect of standardising the name of the county, and to harmonise with the official nomenclature we shall use the name "Breconshire" in the pages of this book.

We have read in one or two sentences the words "County of Brecon." Strictly speaking this is not correct, as Brecon is not a county in the sense that the English counties Kent, Surrey, and Sussex are. Brecon is what is termed a "shire," that is, it was formed from a *share* of a larger district, i.e. the part *shorn* off, for the two words have a common origin in the A.S. *scir*, to cut, to divide. For many years, even after the death of Prince Llewelyn in 1282, the Lordship of Brecknock continued as a Lordship Marcher, and it was only given the privileges that belong to our counties and shires in the reign of Henry VIII. During that reign an Act of Parliament was passed that divided such portions of Wales as had not already been made shire-ground into shires, and of these Breconshire was one. So, to be correct, we should always speak or write of the shire as Breconshire, and not as Brecon as if it were a county.

2. General Characteristics. Position and Natural Conditions.

Breconshire is an inland county—one of the three inland counties of Wales. Its nearest point to the sea is that portion that forms the upper valley of the Tawe, where the boundary line is some 12 miles from Swansea Bay. Crickhowell, the most important place in the south-east, is over 24 miles from the shores of the Severn

Crickhowell Bridge

Estuary, and the rivers Wye and Usk cease to be tidal many miles distant from the borders of the county.

Though Breconshire has not the advantage of a coast-line, nature has not been lacking in the provision of natural attractions that to a certain degree compensate for that loss. Beautiful scenery is found in the many valleys through which the numerous sparkling rivers flow. Its mountains—the highest in South Wales—afford romantic views that are equal to any in their majesty and sublimity. Fish abound in its rivers and lakes, waterfalls of great beauty are frequent, while on the uplands the moors are renowned for their game. Perhaps of all the natural gifts, the medicinal springs at the inland watering places of the county are the most beneficial from a commercial point of view, as they attract large numbers of visitors from other parts of England and Wales during the summer season.

As is but natural in so mountainous a county, the population is sparse, and the chief industry of the inhabitants is agriculture. The mountains are, of course, unsuited for cultivation but make admirable sheep runs, hence a great feature of the Breconshire farm is its sheep. The fertile soil of the valleys, however, amply repays the farmer for its cultivation.

Though agriculture is the chief industry, the county is not without other forms of occupation. The southern border touches the edge of the great South Wales coal-field, thus allowing it to partake somewhat of the prosperity of that important region. The suitability of the district for sheep-breeding and the happy provision by nature of numerous streams, have given rise to a small woollen industry which produces flannel and a coarse woollen cloth. Breconshire also manufactures leather of excellent quality.

Situated in the marchland or border of Wales, with its eastern border contiguous to that of England, it is no wonder that in early days Breconshire was debatable ground. The valleys of the Wye and Usk open to the east, and though these afford ready communication with England in these peaceful times, yet they proved a source of danger and made invasion on the part of marauders coveting its soil an easy matter in more warlike days. This way, no doubt, came the Romans; through these the Saxon hordes made their incursions, and the possession of the valleys of the Wye and the Usk gave Bernard Newmarch and his followers their grip on the county when they rode steel-clad to seek new homes in wild Wales.

3. Size. Shape. Boundaries.

As a rule, when we consider the size of a county, we have to reckon with the fact that counties are divided for purposes of local government into County Boroughs, which administer their own affairs, and the remainder of the county, which is known as the Administrative County. Sometimes, too, a portion of a county is attached to a district in another county for administrative purposes on account of convenience, or a county may administer a portion of another county for the same reason. So the Administrative County does not coincide altogether with the bounds of what is termed the Ancient or Geographical County.

Breconshire has no County Borough, but four parishes on the southern border are within the Administrative County of Monmouthshire, though they lie within the boundaries of the Ancient County of Breconshire. Still, the area thus detached is so small that unless definitely stated otherwise, we shall consider the general term county to mean both the Ancient and Administrative Counties.

Breconshire ranks fourth in size among the counties of Wales and has an area of about 726 square miles or 475,224 statute acres. The area of the Administrative County is 469,301 statute acres. This places it about equal in area to the English counties Surrey and Berkshire and among the dozen smallest counties of the country. It occupies about one-sixtyseventh of the entire area of England and Wales. It is 56 miles long, 35 miles broad, and has a circumference of 140 miles.

The natural formation of the county seems to have marked it off as a mountainous region contained within the bounds of its highlands and streams. In shape it is irregular, but has somewhat the appearance of a slightly truncated right-angled triangle with its base to the south and its vertical side on the west. A smaller irregularly shaped triangle rises from its longest or eastern side.

The boundaries of the county were fixed when the shire was formed in 1535. Certain cantrefs, commots[1],

[1] Brycheiniog at the Survey of Wales in the reign of Howell Dda (tenth century) was partitioned into four cantrefs or cantreds,—i. Cantref

and parishes were named, and of these the new shire was to consist. Being an inland county the boundaries, save when they coincide with a river, are purely arbitrary, but except along the south they are fairly clearly defined. We will now follow these boundaries upon a map, especial care being taken to note when they follow along the line of a river or along a mountain range.

The town of Hay makes a good starting point, as from here the boundary follows the river Wye as far as the northern point of the county. From Hay then the boundary—separating Breconshire from Radnorshire—runs for about seven miles in a south-westerly direction to the vicinity of Three Cocks Junction, when it curves boldly to the north-north-west. This direction it maintains to the neighbourhood of Llanfaredd, on the Radnorshire side, when it takes a sinuous course westward towards Builth Wells. At Builth Wells it runs for about four miles in a north-westerly direction and then turns to the north

Mawr, ii. Cantref Tewdos, iii. Cantref Eudaf, and iv. Cantref Selyf. In the partition of Wales by Prince Llewelyn ap Gruffydd (thirteenth century), as given in the *Myfyrian Archaeology*, Brycheiniog has only two cantrefs, i. Cantref Selyf, comprising the eastern portion of the district, and ii. Cantref Mawr, comprising the western portion. In the list given in Sir John Price's *Description of Wales*, Brycheiniog has three cantrefs and eight commots—

i. Selyf, with two commots, Selyf and Trahayern.
ii. Canol, with three commots, Talgorth, Ystradyw, and Brwynllys or Eglwys Yail.
iii. Mawr, with three commots, Tir Raulff, Llywell, and Cerrig Howel.

When the shire was formed in the reign of Henry VIII, the Cantref of Buallt in Powys was added to the three cantrefs mentioned above to form Breconshire.

with a subsequent inclination to the north-west as far as the confluence of the Elan with the Wye. The boundary now follows the Elan until it reaches the point where the Claerwen flows into the Elan, in the district submerged by the Birmingham Waterworks. It follows the Claerwen valley to the point where the counties of Radnorshire and Cardiganshire meet and then a couple of miles further, near Llyn Gynon, strikes across country southward into the valley of the Towy.

The boundary now lies to the west of the county and follows the Towy valley in a southerly direction until it reaches the spot where the three counties, Breconshire, Cardiganshire, and Carmarthenshire meet. To follow the boundary for some distance further becomes a matter of difficulty. It zigzags now east, now west, now south and sometimes north in seemingly haphazard fashion, but has in the main a south-easterly direction until it reaches the valley of the Gwdderig, when it makes a bend along the western slopes of Mynydd Bwlch y Groes. Leaving this valley it strikes across into the Usk valley, which it follows to the source of the river in the Carmarthen Vans, and crossing that range between Llyn y Fan Fawr and Llyn y Fan Fach strikes south-west into the valley of the Twrch. The Twrch valley is followed until that stream flows into the Tawe near Ystalyfera.

Now the boundary runs along the south of the county and crosses river, valley, and mountain in a manner that baffles description. From Ystalyfera it winds its sinuous way over Mynydd y Drum to Coelbren Junction in the Dulas valley and then over the slopes of Hir Fynydd and

The Blue Pool and Gorge on Taff Fechan

through the valleys of the Pyrddin and the Nedd to Pont Nedd Fechan. Still winding its way eastwards it passes to the north of Hirwaun and then trends over Mynydd Aberdare into the Taff Fawr valley and across Cefn Merthyr into that of the Taff Fechan. From the Taff Fechan valley it crosses Mynydd Llangynidr in a series of straight lines to turn to the south above Brynmawr. From Brynmawr it pursues its winding easterly course, keeping south of Clydach and then bending north it strikes the valley of the Grwyne Fawr. It follows this river valley almost to its head and then striking boldly northwards across the Black Mountains arrives at our starting point—the town of Hay.

4. Surface and General Features.

Breconshire is cradled in mountains, some of which rank as the highest eminences in South Wales. More than half the county is over 1000 feet above the level of the sea, and scarcely any is within 300 feet of it. The general slope of the surface is towards the south and east.

The river valleys divide it naturally into four hill districts. The valley of the Usk, crossing the county from east to west, cuts it into two main portions. North of this valley is Mynydd Eppynt, which is again separated from the wilder hill country of the north-west corner by the valley of the Irfon, a tributary of the Wye. South of the Usk are the Beacons—Bannau Brycheiniog

—which, commencing below the town of Crickhowell, extend westwards into the Forest Fawr or Black Mountains of Carmarthenshire. The remaining hill district, comprising the Black Forest Mountains, lies in the south-eastern corner cut off by the Wye, by the Llyfni, a tributary of the Wye, and by the Usk.

The Beacons, perhaps, are the finest of the mountains in the county, stretching with their spurs along the whole southern border. Three magnificent peaks mark their entry from Carmarthenshire, chief of which Capellante towers some 2394 feet above sea-level. Within the county the peaks succeed each other almost in a line due east and west. Y Gehirrach comes first, 2381 feet high. Then follow Y Fan Nedd (2177 feet), Fan Llia (2071 feet), Fan Fawr (2400 feet), culminating in the peak appropriately named Penyfan, the monarch of the Beacons (2907 feet high), the highest point in South Wales, and the highest Old Red Sandstone peak in the kingdom. East of Penyfan the range slopes downwards again into the valley of the Usk. South of the Beacons, the surface is formed of the high steep barren hills of the great coal basin of South Wales.

Mynydd Eppynt, separating the valley of the Usk from that of the Wye, is most appropriately named, for the range slopes in a long trend in a south-westerly direction from the neighbourhood of Builth. Near the Carmarthenshire border it is known as Mynydd Bwlch y Groes and this spur, continuing round the head waters of the Usk, connects the Black Forest Mountains with Mynydd Eppynt. Though not so high as the Beacons

The Brecon Beacons

—the highest peaks are Moelfre (1450 feet) and Panne (1290 feet)—these hills present a considerable boldness of outline and amongst them lies some of the loveliest scenery of the county. The hills in the north-west are offshoots from the Plynlimmon range, and rise in Y Drygarn Fawr to an elevation of 2120 feet.

East of the valley of the Llyfni—the Vale of Talgarth as it is called—rises the chain of the Black Mountains, or as they are named on the Herefordshire side, the Hatteral Hills. The mountains rise in the well-known Sugar Loaf mountain, near Abergavenny, and range in a convex line with a general north-easterly direction into Herefordshire some distance south of Hay. The highest peaks are Waun Fach (2660 feet) and Pen y Gader Fawr (2624 feet).

5. Rivers—(*a*) Nedd, Taff, Tawe, Rhymney, Sirhowy, Towy.

The rivers of Breconshire are numerous, but though not one of them is navigable, they are yet of importance for the quantity and quality of the fish, especially salmon and trout, that are found in their waters. The river system falls naturally into three divisions—the streams flowing south from the Beacons and their extensions, the rivers of the Usk basin, and those of the basin of the Wye. A glance at the map will show us these divisions and how the Beacons separate the first from the second, and how Mynydd Eppynt forms the barrier between the second and the third.

The streams flowing south from the Beacons do so in a south-westerly and south-easterly direction, the dividing barrier being the huge group of mountains that lies at the head of the Rhondda Valley in Glamorganshire. Just to the west of this barrier flow a number

In the Vale of Neath

of streams that are renowned far and wide for the beauty of the scenery that encircles their head waters. They are the Pyrddin, Nedd Fechan, Hepste, Mellte and Sychnant, which unite their waters near Pont Nedd Fechan to form the Nedd (Neath). Here are innumerable cascades and deeply wooded gorges overhung by craggy heights, varied by woodland nooks of great beauty. This region deserves more than a paragraph and we shall pass on to other streams to return again to a fuller notice of its attractions.

East and west of this romantic neighbourhood are the basins of the two most important rivers of Glamorganshire —the Taff and the Tawe. The Taff rises in two heads, the Taff Fawr and the Taff Fechan, on the slopes of the Beacons. While in the county the Taff is an inconsiderable stream, indeed it is not navigable at any part, but the natural formation of its course has been taken advantage of to construct huge reservoirs to collect the plentitude of water for the use of the citizens of Cardiff. The Taff enters Glamorganshire near the important mining and ironworking town of Merthyr Tydfil, and after winding through a valley containing numerous collieries and ironworks, it flows past the ancient city of Llandaff and enters the Severn Sea at Cardiff.

The Tawe rises in Llyn y Fan Fawr on the eastern slopes of the Black Forest Mountains. It receives numerous tributaries within the county. Chief of these are the Twrch—which for a part of its course marks a portion of the boundary between Breconshire and Carmarthenshire—the Gwysg, and the Giaidd. The Tawe enters Glamorganshire between Ystradgynlais and Ystalyfera, and after a course through a rapidly growing mining and manufacturing district, empties its waters in Swansea Bay.

The Rhymney is another affluent of the Severn that

has its origin in the county. It rises, just within the border, on the Mynydd Llangynidr extension of the Beacons. The Sirhowy has its source in the same mountains, but both streams flow into Monmouthshire before they are swollen by their tributaries to any size. On the western side of Breconshire the Towy, for a portion of its course, forms part of the boundary between the county and Cardiganshire, as we have already seen.

6. Rivers—(*b*) The Usk and its Basin.

The Usk admits no rival to its claim to be *the* river of the county, for it bisects it from west to east and drains the major portion of its area. It rises on the northern slopes of the Carmarthen Van in the Black or Forest Fawr Mountains, five miles south-west of Trecastle. Its course at first is due north and here the river forms part of the county boundary on this side. Some two and a half miles from its source Mynydd Myddfai bars its way and forces it to turn to the east to flow at the foot of the southern slopes of Mynydd Trecastell. At Pont Hydfer it is joined by the Hydfer, which also has its source in the Van, less than a mile from the main stream.

Two miles beyond Trecastle it receives the Crai from the south. This stream rises in Y Gehirrach, and a short distance beyond its confluence with the Usk the latter is joined on the north by the Clydach, which flows from the southern slopes of Mynydd Bwlch y Groes. At Senni Bridge it receives the waters of the Senni whose head waters lie in the valley between Y Gehirrach and Y Fan Llia.

The Cilieni is its next tributary, a very winding stream that rises in a depression between Mynydd Bwlch y Groes and Mynydd Eppynt proper. Near Aberbran station the Usk is joined by the Nant Bran from Mynydd Eppynt. Mynydd Eppynt also is the source of Yr Yscir

On the Senni

(*Outcrop of Old Red Sandstone*)

Fechan and Yr Yscir Fawr, two rushing, tumultuous, mountain torrents that join their waters at Pontfaen to flow as Yr Yscir into the Usk at Aberyscir. On the left bank of the river, on the wedge of land formed by the confluence of the Yscir with the Usk, is the site of Bannium, the military station of the Romans.

At the county town the Usk receives two considerable

tributaries, the Tarrel and the Honddu. The Tarrel rises by two heads, one of which is in Llyn Cwm Llwch, a small glacier-formed lake about a mile in circumference on the northern slopes of Y Fan Fawr. The Honddu flows from Mynydd Eppynt and its junction with the Usk gives to Brecon its Welsh name of Aberhonddu. On Y Gaer Fawr, a height on the left bank of the Honddu,

The Usk near Brecon

are the remains of an ancient fort or camp which gives the height its name, while a short distance south of the camp is the site of Castell Madoc.

The Cynrig and Ogwm flow from the Beacons, and through Glyn Collwng, the valley that lies between the Beacons and Mynydd Llangynidr, flows the Carvanell or Annell, which joins the Usk near Llansantffraed. The

Crawnon is another tributary from the south. It rises in Carn y Bugail in Mynydd Llangynidr and traverses the parish of that name.

Only two tributaries of any size flow from the Black Mountains—the Rhiangoll and the Grwyne. Between the confluence of these two rivers with the Usk lies Crickhowell, the second town of any importance that stands on the Usk while a Breconshire river. The Rhiangoll rises in the Gader ridge and flowing past Cwmdu and Tretower joins the main stream at Glanusk Bridge. The Grwyne rises in two heads—Grwyne Fawr and Grwyne Fechan—on opposite sides of the same Gader ridge. They join their waters a little south of Llanbedr and flow into the Usk half a mile south of the village of Llangrwyne. The Usk now ceases to be a Breconshire river as it flows into Monmouthshire a little west of its confluence with the Grwyne. It winds its way past Abergavenny, and flowing in a south-westerly direction, meanders peacefully through the town of Usk to Newport and the Severn Sea.

7. Rivers—(*c*) The Wye and its Basin.

The Wye rises about a mile eastwards of the highest summit of the Plynlimmon range, in a district hallowed by historic memories. Around its upper reaches waged many a stern conflict in days of old. Owain Cyfeiliog, Prince of Powys, here met and fought Howell ap Cadwgan, and here also Owain Glyndwr unfurled the banner of Welsh independence. In Plynlimmon, Glyndwr had his headquarters in the summer of 1401 and from thence set out with his forces to harass the lands held by his enemies, burning Welshpool, sacking Montgomery, and destroying the Cistercian abbey of Cwm Hir before he returned.

The Wye at Builth

Two miles from its source the Wye is joined by the Tarenig, a mountain stream held by some to be the true fountain head of the river. It flows through a wild and barren district until it reaches Llangurig. Here it takes a bend to the south and the scenery on its banks becomes of a richer and bolder character. The Radnorshire town of Rhayader Gwy is passed and, shortly after, the river enters Breconshire. From its entry into our county, until it leaves it at the town of Hay, the Wye flows through scenery of the most beautiful description, thought by some to be unrivalled among the inland river scenery of the country, especially at the fall of the year, when the warm

The Irfon in flood

tints of autumn-dyed leaves surround the river with a wealth of colour.

Where the Wye enters Breconshire, it receives on its right bank the waters of the river Elan, which, with a tributary the Claerwen, forms the northern boundary at this point. Cwm Elan, transformed into a huge lake, forms the reservoir whence the great city of Birmingham derives its water-supply. At Builth the Irfon, a beautiful and important stream, rising on the slopes of Bryn Garw in the north-western corner of the county, adds its waters to the Wye. Its waters are fed by many mountain streams from the wild hills around, chief of which are the Cerdin, Camddwr, Cammarch, Dulas, and Chwerfu on the left bank, and the Dulais on the right.

The valley of the Irfon is fairly well populated, many villages and towns being found on its banks. The presence of mineral wells has helped to increase the prosperity of two of these, Llanwrtyd Wells and Llangammarch, which are fast becoming favourite summer resorts. Below Builth only two streams of importance fall into the Wye. These are the Dihonw and the Llyfni. The Llyfni flows from Llyn Safaddan or Llangorse Lake, and on its banks are Trefecca and Talgarth. Hay lies some five miles beyond the junction of the Llyfni and the Wye and just beyond the town the Wye ceases to be a Breconshire river and flows into the county of Hereford.

8. Lakes and Waterfalls.

For so mountainous and well-watered a county, Breconshire has few lakes. In the north are Llyn Ffarlyn, which lies among the slopes of Pen-y-Gorllwyn, and Llyn Carw, on Cerrig Llwyd y Rhestyr, an offshoot of Y Drygarn Fawr. Among the hills of Mynydd Eppynt are Llyn Llogin and Llyn Penylan, and above Llyswen on the Wye is Llyn Brechfa. These are but small sheets of water hardly worthy of the name lake.

South of the Usk, however, are two lakes more deserving of the term—Llyn y Fan Fawr, from which the river Tawe flows, and Llangorse Lake, the largest natural sheet of water in South Wales. Llyn Safaddan or Safeddan, as the Llangorse Lake is named in Welsh, has a circumference of about five miles; its greatest length is two miles from east to west, and its greatest breadth is about one mile. Its waters abound in fish of several kinds, chiefly pike, roach, and perch, and eels of such an extraordinary size as to have given rise to the proverb, "Cyhyd a llysywen Safaddan" (as long as a Safaddan eel). The lake is the subject of many fables and traditions, and at one end of it archaeologists have discovered traces of the lake-dwellings of a prehistoric race of men. It is said that the waters of the Llyfni do not mix with those of the lake and that the fish from the one never pass into the other.

Brief reference has already been made to the natural attractions of the south-western corner of the county, where the waters of several streams unite to form the Nedd. The limestone of this district is easily worn away by the action of water, and as the strata above it are generally of sandstone, which better resists the action of water, the wearing away of the one and the resistance of the other give rise to those precipitous step-like formations over which the waters tumble in many beautiful forms. The action of surface waters often eats away the rock into huge pits, and several of these exist on the mountain slopes of this district.

The Pyrddin lies farthest west of the five streams

and before it joins its waters to the Nedd Fechan, it leaps in two beautiful cascades from shelf-like sandstone rocks into the limestone basin below. Scwd Einion Gam is the first. The river approaches the fall through a gloomy, narrow glen, bordered with mountain ash and willow, and plunges a distance of from 70 to 80 feet in one grand unbroken sheet into the basin beneath. Three-quarters of a mile below is Scwd Gwladys, or The Lady Fall, where the water has a fall of 40 feet and shoots clear of the ledge beneath so that one can get under it though he cannot cross to the other bank.

The Nedd Fechan also has two falls, the Upper and the Lower Falls. These are pretty, but neither are they so high nor is their environment so picturesque as that of the other falls of the district. On the Mellte are three falls, the Upper, Middle, and Lower Clungwyn Falls. The Lower Clungwyn Fall is about 40 feet high. It is fine and massive, as even after a drought a considerable amount of water is found in the stream. A short distance above is the Middle Fall, which is more distributed, and descends from a curved rock suggestive in miniature of the form of Niagara.

The Lower Hepste Fall on the river of the same name is really a steep cataract, as the water tumbles down in a series of mad leaps, but the Cilhepste Fall, higher up the stream, is the most beautiful waterfall in the district. The valley in which it is situated is deep and almost inaccessible, and through it the water leaps in a wide unbroken sheet from a level rock nearly 50 feet in height into a deep basin. Ysgwd yr Eira, as it is

Upper Ffrwdgrech Waterfalls

sometimes called, throws its waters clear of the rock and, by means of a ledge about 3 feet in width, a passage is possible from one bank to the other underneath the falling sheet of water. This is the only path from one side of the valley to the other. The glen is full of mist, on which the sunlight plays at times with beautiful rainbow effects.

Before leaving this neighbourhood we must notice some other curious natural features, the underground cavern through which the Mellte flows, Craig y Ddinas, and the wild gorge of the Sychnant or Sychryd. The underground cavern, "Porth yr Ogof," lies some distance south of the village of Ystradfellte and through it the river Mellte flows for over six hundred yards. The mouth of the cavern is very picturesque, surrounded as it is by the vivid greenery of the glen. A broad slab of rock forms the lintel to the entrance, which is some 20 feet in height and 45 feet in width. Inside, the cavern opens out into a large apartment overhung by numerous stalactites. At no time is the cavern accessible from end to end, for in its passage the river falls with a tremendous roar into a deep part at a lower level, but when the water is low it may be traversed until light is visible at the lower end. The subterranean passage of a stream is quite characteristic of a limestone district, as the friction and chemical action of the water on the lime during the course of ages wear away a passage through the rock which is ever slowly increasing in extent. A rock of different composition would resist the action of the water and cause the stream to take a different course.

Parallel with the Hepste flows a small brook called Y Sychnant, which here forms a portion of the boundary between Breconshire and Glamorganshire. The Sychnant flows into the Mellte, and on the tongue of land formed by their confluence stands a bold precipitous limestone rock of considerable elevation known as Craig y Ddinas. Its top is fairly level, and is accessible after a struggle, when the climber is rewarded with a magnificent view of the Vale of Neath and the Mellte glen. The ravine of the Sychnant is particularly fine, but difficult of access. The river is enclosed by tall cliffs and its bed is a cradle of boulders. Except during floods the brook is subterranean for a portion of its course through the glen, but in flood the rush of waters forms a series of cataracts that give to the gorge its name of "The Cascade."

Upper Ffrwdgrech Waterfalls on the Rhyd Goch river also show glen scenery characteristic of the district, and Scwd yr Hen Rhyd is another waterfall that deserves notice. The Llech, a tributary of the Tawe, a short distance north of Coelbren Junction, falls in one unbroken sheet of water, some 90 feet perpendicularly. Pistyll Mawr on the river Clydach is another fine waterfall, and there are many others of greater or less height and beauty on the several streams of the county. It is, indeed, a land of waterfalls.

9. Geology and Soil.

The science of Geology aims at an investigation of the structure of the earth, and explains how the different varieties of rocks have been formed, and how they have been arranged in the positions they occupy. In short, by Geology we mean the study of the rocks of the earth's crust, including in the term "rock" all mineral substances, hard or soft, even to loose sand and soft clay.

Rocks are divided into two broad classes, those formed by the action of fire or heat—termed Igneous Rocks—and those laid down by water—termed Aqueous Rocks—or, as they are found arranged in sheets or beds—Stratified Rocks. Water-formed rocks were once carried in suspension by the streams that were responsible for their existence and the materials held by them in suspension were deposited as a sediment, hence an alternative name for this class of rock is "Sedimentary Rock."

Igneous rocks were once in a molten state and assumed their solid form on cooling. The oldest rocks in the world were so formed, and on them the later stratified rocks were laid. Still we must not consider the igneous rocks now visible as part of the original rocks of the world. These have been formed at a later period and are often much younger than the stratified rocks among which they lie. Wherever there is volcanic action they are still in process of formation. Lava from volcanoes flows over the surface or is forced among fissures in other rocks, and cooling,

	Names of Systems	Characters of Rocks
TERTIARY	Recent & Pleistocene	
	Pliocene	sands, superficial deposits
	Eocene	clays and sands chiefly
SECONDARY	Cretaceous	chalk at top sandstones, mud and clays below
	Jurassic	shales, sandstones and oolitic limestones
	Triassic	red sandstones and marls, gypsum and salt
PRIMARY	Permian	red sandstones & magnesian limestone
	Carboniferous	sandstones, shales and coals at top sandstones in middle limestone and shales below
	Devonian	red sandstones, shales, slates and limestones
	Silurian	sandstones and shales thin limestones
	Ordovician	shales, slates, sandstones and thin limestones
	Cambrian	slates and sandstones
	Pre-Cambrian	sandstones, slates and volcanic rocks

forms a new igneous rock in those places. Again volcanic ash or dust is thrown up by volcanoes and oftentimes covers a considerable area in a thick layer.

The oldest stratified rocks were formed by the disintegration or wearing down of the primitive or first-formed igneous rocks by climatic or chemical influences. Running water collected the particles and carried them into the beds of lakes or into the sea, and matters held in solution were also deposited by the waters in course of time. Thus the various beds of sandstone were deposited; thus the clay beds that were afterwards heated and compressed to form slate were made; and thus beds of rock salt, of gypsum, and of carbonate of lime were deposited by the waters that contained these substances in solution.

These beds suffered many changes at times. The movements of the earth would lift them up or depress them, or again other beds would be deposited over them and they would be subjected to pressure and heat. The earth movements formed the beds into arches or troughs in various ways so that they would be contorted out of the flat or nearly flat position in which they were originally laid. If we take a bundle of tailor's patterns and press it at both ends we see something like the condition into which most of the rocks of the earth's crust have been twisted. Apply pressure above and below and you will obtain more varied forms still. Often some of the beds would break under the strain and one portion be forced above or below the level of the other, giving rise to what geologists term a "fault."

The pressure exerted upon rocks changed their nature.

Bwa Maen

(Contorted Limestone)

They were hardened so that a bed of sand would become hard sandstone, soft deposits of lime became limestone, muds and clays became shale, and so on. Then the combined pressures on all sides caused joints or cracks : the side pressure often causing the rocks to split easily into sheets of varying thickness. This result—termed "cleav-

Striated Boulder

(*Marked by glacial action*)

age"—is best seen in slate, which can be split into very thin sheets.

When rocks are tilted and folded by the movements of the earth's crust, the weather, rivers, glaciers, or waves of the sea wear away the crest of the curve just as one might cut the top of the curve in the cloth patterns with a pair of scissors. Thus we often see rocks cropping out at an angle, with the edges of the strata showing on the surface. Perhaps on this broken edge another bed of rock may be deposited, and sometimes a comparatively recent rock may be seen lying on the top of a much older one, the causes just enumerated having taken away one or more layers of the rocks that should lie between. Or again glaciers may transport boulders which have fallen upon them, and after travelling some distance deposit them as isolated rocks on recent strata.

The arrangement of the beds is important, as by their positions geologists are able to classify the rocks into groups according to their age, the upper beds being of course newer than those that lie beneath. The stratified rocks are divided into three main groups called Palaeozoic or Primary, Mesozoic or Secondary, and Cainozoic or Tertiary. Below the Primary rocks are those termed pre-Cambrian, and these form, as it were, the foundation on which all the other rocks have been built.

It is a curious circumstance that in Great Britain we find all the older rocks in the north and west of the island and the younger rocks in the south and east. A line drawn from the Exe to Whitby roughly separates these two marked geological divisions. The reason is not far to seek when we consider that during the submerged period we shall read about later, the lower-lying eastern and southern portions were more favourable to the formation of new sedimentary rocks than the more rugged and elevated northern and western portions.

An examination of a geological map of Great Britain

will show us that the Cainozoic or Tertiary rocks, in the form of sands and clays, lie principally in the Thames Basin and in a part of Hampshire. They overlie the chalk of the Mesozoic or Secondary period, which out-

Glacial Boulder

(*Granite on Silurian rock*)

crops farther to the west and north. West and north of the chalk appears the Oolitic series, which in turn gives way to the Lias formations. The Lias is succeeded by the marls and new red sandstone of the Trias formation, which pass into the Palaeozoic or Primary rocks of the north and west. These last occupy nearly two-thirds of the surface of Britain.

Now that we have a general and rough idea of the classification of rocks we can proceed to a brief notice of the rocks found within the borders of Breconshire.

All the rocks of the county are of Palaeozoic age and a greater portion of its surface is made up of the Silurian and Devonian formations of that great group. The Lower Silurian rocks are found in the north-west of the county in the district extending from the Wye to Llanwrtyd. The lowest layers of this formation consist of a series of strata called Lingula Flags, which are made up of dark shales and flagstones. Next to them come the Llandeilo Beds, so called from Llandeilo in Carmarthenshire, which are composed of dark slates and sandy flagstones with occasional beds of sandstone. Other beds of the same system occur, and the characteristic nature of the rocks is that they are shales and sandstones with occasional thin layers of limestone.

The Lower Silurian rocks are overlaid by those of the Upper Silurian formation, which pass through the Tilestones, forming the upper bed of that system, into the Devonian rocks that lie to the south of a line drawn from Llanwrtyd to the Wye. The chief rock in the Devonian system is the Old Red Sandstone, which occupies the central and most of the southern portion of the county. It is of this rock that the higher mountains such as the Fans, the Beacons, and the Black Mountains are composed.

The southern boundary of Breconshire touches the

edge of the South Wales coalfield. Here the Old Red Sandstone passes below the Carboniferous rocks. The first or lowest layer of this system consists of a narrow pan or layer of Carboniferous or Coal-bearing Limestone. This is succeeded by a pan of hard sandstone called Mill-stone Grit. When South Wales miners strike this rock they know that they need not dig deeper in search of coal, and for the reason that no coal has ever been found beneath it, they call it the "Farewell Rock." Carboniferous Limestone is also seen on the hill called Pen y Carreg Calch, between Cwmdu and Llanbedr Ystrad. It has been suggested that this forms the remains of a "great limestone coat of mail, surmounted by coal measures and beds...which, removed by stupendous water action, left this capstone behind."

The geological map shows us that north of Llanwrtyd Wells lies a narrow strip of that form of igneous rock known as Andesite. In some long-past day a volcanic disturbance forced the molten lava almost to the surface, and in course of time the erosion of the rocks that once covered it has brought the hardened rock to the surface. As the county is entirely inland there are no deposits of sand, but, in the district near Hay, the Wye has deposited some quantities of alluvium.

The county contains a great variety of soils on account of the various rocks upon which and from which they have been formed. In the northern part, included in the slate, or rather shale tract of South Wales (Silurian), we find peat, as a rule, occupying the hollows and sometimes the slopes of the hills. Here the sub-soil is clay, which renders the ground wet and unproductive of any but the poorest herbage. The banks of the Wye and Irfon, nevertheless,

Outcrop of Silurian Rock

are composed of land of a much richer quality, where a sound loam prevails to the depth of from one to six feet.

The soils of the Carboniferous district are for the

most part of the same poor quality as those of the uplands of the slate district. A clay sub-soil makes them wet. It is said that, owing to their containing a fine silex which renders them friable, the clayey soils of this district are more capable of improvement under cultivation than those of the north. The use of lime makes a soil of this nature sweeter and drier, and fortunately there is plenty of this mineral in the district.

The middle part of the county is wholly composed of soil having the Old Red Sandstone as substratum. For the most part it is composed of a brownish-red sandy loam, which in the Vale of Usk is highly sandy in nature. This soil is light, and during dry summers does not retain sufficient moisture to nourish the plants grown in it. When well cultivated and irrigated it returns fairly good crops. The soil of the uplands—as distinct from the mountain tracts—is stronger, having more argil in its composition, and under good tillage it produces good crops of grain. The Vale of the Wye below the town of Builth has brown gravelly loam, capable of growing fairly fine crops, but near Glasbury, where the river widens and the river deposits are composed of well-mixed particles from both shale and sandy districts, and in the Vale of Talgarth, the soil becomes a fine rich loam. The mountain tracts of the Red Sandstone districts are far superior in soil to the corresponding tracts in the shale district. The soil of the narrow limestone district between the red soil and the coal measures is for the most part rendered very arid by its elevation, its want of depth, and the absorbent quality of the sub-soil.

10. Natural History.

Thousands of years ago the British Islands formed part of the mainland of the Continent of Europe, and it was but the other day, as time is reckoned by geologists, that they became separated and were made islands. Many things prove that such was the case, as, for example, the existence of the remains of forests now sunk beneath the sea, the comparative shallowness of the waters surrounding our coasts, and the similarity in construction of our shores to those of the continent directly facing them.

Geologists also tell us that before these islands were severed from the mainland, they were, at one period, almost entirely submerged by the sea—as the existence of marine shells on the summit of one of the Carnarvonshire mountains goes to prove. Of course such a condition was fatal to the animal and vegetable life of the land, so, when it again appeared above the surface of the waters, it was necessary that it should be re-stocked from a territory where plants and animals were already in existence. This was the mainland of Europe, and as our land was connected with the Continent on the east and south, it, in course of time, became covered with vegetation, which, naturally, was followed by an influx of animal life. So we must expect to find that our flora and fauna are similar to those of the Continent; and this, with slight modification, is the case.

Some of the land failed to keep above water and sank,

or was washed away, forming the North Sea and the English Channel, and as this occurred not very long after the period of submersion, all the continental species could not establish themselves within our borders. In fact, even at the present time, the districts nearest the Continent, as the south and east of England, are richer in the number of their species than are Scotland, the west of England, and Wales. Ireland, the remotest part, is poorer in flora and fauna than almost any part of Britain.

Breconshire, in comparison with the greater portion of England and Wales, is not rich in flora for the reason stated above, and also because its surface as a whole is elevated and sterile. Taking its flora first, we shall only notice a few of the rarer plants or those that are of special interest. In the latter class we may include the daffodil, which is plentiful in the Wye valley, especially in the neighbourhood of Builth. Chives (*Allium schoenoprasum*) are common on the banks of the Wye, and in one or two districts the root-parasite, *Lathraea squamaria*, is common. The Welsh poppy (*Meconopsis cambrica*) is fairly abundant in some parts. Among the rarer plants of the county may be mentioned the ivy-leaved bell-flower (*Wahlenbergia hederacea*), spreading bell-flower (*Campanula patula*), broad-leaved bell-flower (*C. latifolia*), Herb Paris (*Paris quadrifolia*), Grass of Parnassus (*Parnassia palustris*), maiden pink (*Dianthus deltoides*), mossy saxifrage (*Saxifraga hypnoides*), globe flower (*Trollius europaeus*), mountain pansy (*Viola lutea*), vernal sandwort (*Alsine verna*), and blue pimpernel (*Anagallis caerulea*). The moist moors and heaths are the habitat of the needle whin (*Genista anglica*), and in the bogs both the round and long leaved sundews (*Drosera*) are to be found. In damp pastures the meadow plume thistle (*Carduus pratensis*) occurs, and the melancholy thistle (*Cnicus heterophyllus*) is found in sub-alpine pastures and rivulets. In some of the swampy districts the graceful fronds and inflorescence of the royal fern (*Osmunda regalis*) wave above the surrounding herbage.

The number and species of the mammals are those of almost every inland county in the country. Badgers are common and are said to be increasing in numbers. Foxes, too, are numerous, and otters are fairly so. The stoat and weasel are frequently seen; the polecat makes its appearance occasionally and the pine marten is not yet quite extinct. The pipistrelle, whiskered noctule, and long-eared bats are fairly common. The barbastelle bat was observed in the porch of Llanelwedd church (Radnorshire) in 1904, but before this date its presence had not been recorded west of Worcester. Though Llanelwedd is so close to our border this bat has not yet been recorded in the county.

Reptiles are few in number. The viper or adder is but very occasionally seen and so is the common snake. The lizard and slow-worm are fairly numerous.

The Usk and Wye are among the best known salmon rivers in the country and these rivers and their tributaries abound in trout. The reservoirs in the county are also well stocked with trout. Chub and dace are plentiful in the rivers and pike are always to be found. A very local fish in the Wye is the allice shad (*Clupea alosa*), which may be seen in shoals in May and June. Llangorse

Lake, too, forms a favourite resort of anglers, pike, perch, and eels of large size being the chief fish caught. The canal swarms with roach.

Breconshire, a county of mountains, moors, and river valleys, of mountain tarn and bogs, presents diversified physical features each suited to the needs of different species of birds, and we naturally expect that its list of birds should be varied and extensive. Though nearly 200 species of British birds have been recorded as occurring in the county many of these are but migrants, visitors and often storm-driven stragglers. Still over 100 species may be classed as residents, making the county their home during the breeding season.

The larger land birds are fast becoming extinct. The last golden eagle was shot some 50 years ago; the osprey, once to be found on Llangorse Lake, has not been seen for many years; the marsh-harrier, formerly common in the south of the county, is now extinct; and the hen-harrier, though occasionally seen, has almost left the county. The buzzard still holds its ground and is fairly common in the rocky hills of the north and in the wildest recesses of the Beacons, and the honey buzzard sometimes makes its appearance in the south-west. The kite remains and has been known to breed within recent years, the peregrine falcon is frequently observed, and the merlin and hobby are occasionally seen. Sparrow-hawks and kestrels are common. The brown owl still maintains its position, though the short-eared, long-eared, and white or barn owls have almost all disappeared.

Among the rare winter visitors are the golden oriole

and the great grey shrike. The ring ouzel, a bird of the uplands, is a frequent summer visitor, breeding sometimes on Mynydd Eppynt, and the clear mountain streams are the haunts of the water ouzel or dipper—one of the commonest of our birds.

The remote regions of the county still afford shelter to the raven: the hooded crow is becoming rare. Starlings breed in large numbers and during the autumn their numbers are augmented by large flocks from other parts which seem to have a striking partiality for the reed beds of Llangorse Lake. These also attract great numbers of reed and sedge warblers. The garden and grasshopper warblers occur but rarely, though they are said to be increasing. The nightingale occurs sparingly, and it is said locally, with some foundation in fact, that it is never heard west of Bwlch, eight miles south-east of Brecon. The many fir and larch plantations of the county form the haunts of numbers of golden crested wrens.

The coal titmouse is the rarest of the commoner tits, the great, blue, and long-tailed species being numerous. The fact that the Usk is a good trout stream no doubt accounts for the numerous wagtails that are residents, the grey headed wagtail being the only uncommon member of that family. The woodlark is also a rare bird, but meadow and tree pipits have increased considerably of late years.

The twite is a fairly common winter visitor, being distributed throughout the county. The tree sparrow occurs rarely, as does the hawfinch, though the latter is increasing and has taken to breeding in the district around

Llangorse Lake

Crickhowell. A rare winter visitor is the crossbill, though it occasionally occurs in large flocks attracted, no doubt, by the numerous coniferous trees of the county. The wryneck is very rare but has been known to nest in the county. The kingfisher is found in slowly increasing numbers on the larger streams, and the night-jar is fairly common on the lower hills.

The heather-covered moors of the county form the home of grouse. The black grouse is said to be native and, being strictly protected, is increasing in numbers. The red grouse is also plentiful and increasing, the grey partridge is common but the red-legged partridge is very rare. The quail is an occasional visitor, and some years ago a pair nested near Brecon but did not hatch. The bogs among the hills are the haunts of the great snipe, and the common and jack snipes. The swampy districts too form the breeding grounds of the curlew.

The golden plover occasionally breeds on Mynydd Eppynt, and Llangorse Lake is sometimes visited by the ringed plover and the turnstone. The greenshank and the common redshank are very rare in the county though both species have been observed and shot in the vicinity of Llangorse Lake, and the latter also on Mynydd Eppynt. The green sandpiper is also very rare, although there is some evidence of its having bred in the county, and solitary specimens of the curlew and purple sandpipers have been recorded. The common sandpiper is frequent, breeding on the banks of the Usk and the Wye, which rivers with Llangorse Lake form the occasional haunts of the dunlin.

The heron is generally distributed over the county, but as there are few nesting places and no well established heronry many of the birds must have been bred elsewhere. Though once common the bittern has become scarce, Llangorse Lake and the Usk valley being the localities it visits. The swampy places are the haunts of the water-rail, which is fairly common, and to the bogs in the vicinity of Onllwyn and Hay the spotted crake is a regular visitor.

Llangorse Lake seems to be the favourite haunt of the swimming birds, though some, as the moorhen, also frequent the lesser lakes and the pools of the Usk and Wye. The coot breeds on Llangorse Lake and also on Gludy Lake near Brecon. The sheldrake, shoveller, and pintail occur very rarely, but the wild duck breeds in the bogs in the hills, is fairly common and is increasing. The wigeon is also a common winter visitor and the teal breeds in several bogs among the hills. A winter visitor that occurs frequently is the goosander. The little grebe is common and breeds in several parts, and the great crested grebe breeds regularly on Llangorse Lake.

11. Climate and Rainfall.

By climate we mean the kind of weather that a country, county, or district enjoys throughout the year. The nature of the climate of a particular district depends upon a variety of conditions, chiefly, the temperature of the air, the direction of the prevailing winds, the amount of moisture in the air, and to a lesser degree, the character of the soil and the nature of the surface.

Perhaps the most important factor in determining the temperature of the air is the distance of the district under consideration from the Equator. Breconshire is situated about midway between the Equator and the Pole, and thus enjoys what is known as a Temperate climate, that is, it is not subject to the excessive heat of the Tropics, or to the extreme cold of the Polar regions.

Another factor of importance is the proximity of the sea. Nearness to the sea modifies the climate of a district to a considerable extent, and, though our county is entirely inland, it is still near enough to the Atlantic Ocean to benefit from the cooling breezes of summer and the tempering winds of winter. Again the surface of the county is hilly, and in fact more than half of the surface of Breconshire is over 1000 feet above the level of the sea, and scarcely any of it is within 300 feet of it. This high altitude makes the temperature colder than if the county lay at a lower level, and scientists have computed that an average fall of one degree Fahrenheit takes place with every 270 feet of ascent above the sea-level.

The prevailing winds come from the west and the south-west, thus making the climate a moist one. These winds come from the Atlantic Ocean laden with moisture, and striking the mountains are driven into higher altitudes where the water vapour is condensed to fall as rain. A glance at the rain chart will show that the wettest regions lie in the south and west and that to the eastward and northward the rainfall gets less.

The nature of the climate is of great importance, as it has considerable influence upon the vegetable and animal productions of the county. We have established the fact that the climate of Breconshire is, generally, mild and moist, and such a climate favours the production of those plants whose leaves are of more service than the fruit. Thus, apart from the quality of the soil and the great average altitude of the county, the climate of itself, except in portions of the valleys, prevents any successful growth of grain. With reference to animals we shall take the sheep as an example. We find that in damp climates the wool does not reach the same quality as in a drier district, and this fact is noticeable even within the confines of such a small area as Breconshire. On the other hand, the mutton from the wet districts is superior to that from the drier parts.

A farmer should always have a knowledge of the climatic influences at work in his neighbourhood so that he may be guided in the selection of his flocks and in the crops suited to his district. Of considerable service to this end, among other things, are the stations established in different parts of the country (there are 4000 such stations in the British Isles) where particulars of the rainfall, temperature of the air, the force and direction of the winds, barometric pressure, and the number of hours of sunshine are collected daily and tabulated. Such observations are sometimes published daily in the local press and they are all sent to a central station, the office of the Meteorological Society in London, where charts are prepared giving the information collected in a complete and graphic manner.

From a careful study of cause and effect and of the statistics furnished by these returns, the society's officials are enabled to forecast, with a considerable degree of accuracy, the nature of the probable weather for the next twenty-four hours in the 12 districts into which the British Isles and the Western Channel are divided. Warnings are also issued when rough weather is expected.

Not only do we get these daily particulars, but they are also tabulated for the various stations and issued in book form, with charts. From these we learn that in the Beacons district from 60 to 80 inches of rain fall annually. Just beyond the border, in Glamorganshire, over 80 inches fall in the year. South of the Usk the average annual rainfall decreases to a range of from 50 to 60 inches, while the same figures hold good for a small district in the north-west of the county. Between these two districts lies a region where the average ranges from 40 to 50 inches yearly, and along the right bank of the Wye, from Builth to near Hay, there is a narrow strip where the average varies from 30 to 40 inches a year. In 1908 the wettest portion of the county was around Bwlch, where the instruments recorded the high fall of 135 inches of rain.

A comparison of the annual rainfall of our county with that of Great Britain as a whole shows us that we experience the heavy rainfall that is the common lot of the western and mountainous areas of the island. A portion of the county falls in the area of greatest rainfall. But when we examine the records for the eastern portion

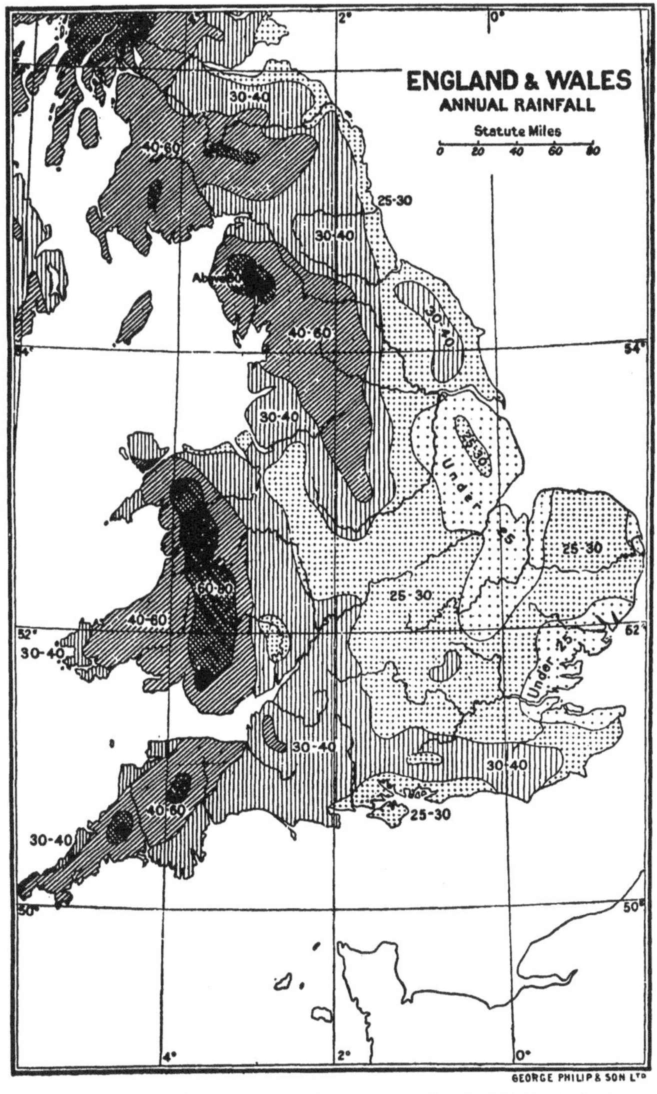

(*The figures give the approximate annual rainfall in inches.*)

of the island the contrast is very marked; as in the east, instead of a rainfall varying from 30 inches per annum to 80 inches and over, it varies from 25 to 40 inches only in the wettest parts, with an area lying east of the Trent basin, in the basins of the Wash rivers, and surrounding the Estuary of the Thames, where the annual rainfall is under 25 inches.

In many places records are kept of the number of hours of bright sunshine enjoyed at those places. From these records we find that the regions of less rainfall—that is the east and south coasts—have also the greater number of hours of sunshine during the year. The total amount in the sunniest districts is about 1800 hours, which decreases the farther north and west we get to about 1200 hours per annum. There is only one station in Breconshire at which sunshine records are kept, that maintained at Llangammarch Wells by Dr W. Black Jones, who very kindly supplied the statistics for this paragraph. In 1910 the total amount of bright sunshine recorded at Llangammarch Wells was 1246·3 hours. The sunniest months were May, June, and July with 206·2, 161·5, and 155·5 hours respectively. August was very wet but 112·0 hours of sunshine were recorded. Other fairly sunny months were March with 136·0 hours and September with 118·1 hours. The least sunny months were December with only 16·3 hours, January with 45·1 hours, and November with 58·7 hours. The average sunshine per day for the year amounted to 3·4 hours. The year 1910 was an exceptionally wet year for the district, so taking this fact into consideration the district around Llangammarch Wells enjoys a very fair amount of sunshine for a place situated among the mountains.

A chart giving the mean or average temperatures (isotherms) in July and January shows us that in July Breconshire is contained within the lines showing a mean temperature of 61° and 62° Fahr. These lines run from south-west to north-east, and the warmest portion of the county is the south-eastern corner which lies farthest from the cooling influences of the sea. The lines for January cross those for July almost at right angles and run in a direction nearly north and south. They show a mean temperature ranging from 40° to 30° Fahr., and now the warmest part of the county is the south-west corner, the one that lies nearest to the sea and feels with the greatest force the tempering influences of the breezes warmed by passing over the North Atlantic Drift that sweeps through the ocean towards these shores.

12. People—Race, Language, Population.

When Britain was still joined to the Continent of Europe it was the home of a race known to us as the Palaeolithic—men, that is, of the Old Stone Age. They dwelt in caves in the earth or in lake dwellings, and they were in a very low state of civilisation. So far as we know they had no domestic animals; they did not till the soil; they lived mainly on the flesh of animals which they hunted. Their tools and weapons were of the rudest description, chiefly made of roughly chipped flints. This race of people was followed—after a long period during which the land disappeared to give place to the English Channel—by the long-skulled, dark-eyed Iver-nians who are sometimes known as the Neolithic or New Stone Men. These were more cultured than the Old Stone Men and they had learnt how to make finely

Prehistoric Implements found in Breconshire

polished and ground implements and weapons of stone; they possessed domestic animals, and they knew how to make rough cloth and rude pottery. They lived in long huts and buried their dead under long mounds. After another long interval came the first division of a Celtic people known as the Goidels, who were followed by another branch of the same race known as the Brythons. These Celts were round-skulled, had blue eyes and fair hair, and made weapons and implements of iron and bronze.

They lived in circular huts and burned their dead before burying them under round mounds. The Silures, who inhabited Breconshire when the Romans came into the country, were of the Goidelic branch of the Celtic race,

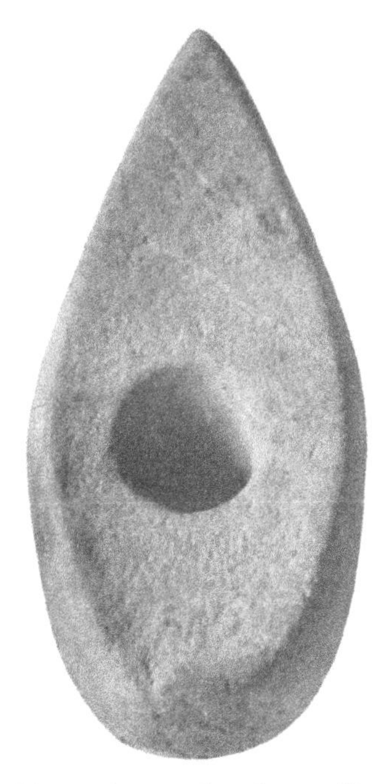

Neolithic Implement found near Devynock
(*Welsh Museum, Cardiff*)

but the Brythons had driven a wedge right across mid-Wales, so it is possible that some Brythons had overflowed into the northern part of the county.

As each new wave of people arrived on the shores of Britain they dispossessed their predecessors of the best lands and in time drove them westwards and northwards into the more mountainous and barren regions. Many of the inhabitants were doubtless enslaved by their conquerors, to become, in process of time, assimilated into their race. The Goidel dispossessed the Ivernian, and, as we have seen, the Brython in parts took the place of the Goidel.

The Romans ruled the country for nearly four hundred years, but in spite of their long residence they do not seem to have left any permanent racial trace on the people. During the last few years of the Roman occupation bands of Irish Celts made settlements in Wales, but they do not appear to have settled any portion of this county. Shortly after the Roman exodus a Brythonic force from Strathclyde, under Cunedda, subdued these intruders and established itself in their place. We are told that these North Brythons came in three waves, that a great portion of the country was settled by them, and that the family of Cunedda gave us our ruling families. Under Cunedda the Celtic peoples formed themselves into a great confederation and henceforward they are known by the generic title of Cymry. From this time the old tribal divisions of Silures, Ordovices, and so on are lost sight of, and the country appears to have been divided into a number of small principalities, which were to all intents and purposes independent, though nominally under the sovereignty of an overlord who was a descendant of Cunedda.

The Saxons made repeated incursions into Breconshire, and though evidences exist of mounds erected by them on which their strongholds were built, they left no trace on the people. The Normans, however, left distinct traces. They conquered and settled the district, and by marriage and inter-marriage a large infusion of Norman blood resulted.

Thus we see that the people of Breconshire who have descended from Breconshire forefathers for some generations are the sons and daughters of the prehistoric people, the Goidelic Silures, with some traces of the Brythonic Ordovices and the Brythons from the north, and of the Normans who came, *via* England, from Normandy.

We must now turn our attention to the people of Breconshire of to-day. We get our particulars from the census returns that have been taken every ten years since 1801. In 1901, 59,907 persons inhabited the Ancient County. Of these, 38,515 were born in Breconshire, and 13,302 were born in the other counties of Wales and Monmouthshire. No less than 476 were born in London, 204 in Scotland, 278 in Ireland, 43 in the Colonies, 7 in the Isle of Man and the Channel Islands, and 98 were subjects of foreign nations. The remainder of the population was born in the English counties, the West Midlands (3600), and the South Western counties (1200)—the counties lying nearest to Breconshire—providing the majority[1].

According to the 1911 census, 59,298 persons inhabited the Administrative County, an increase on the

[1] Similar details of the 1911 census are not yet published.

54,213 inhabitants in 1901 of 9·4 per cent. This makes an average of about 81 persons per square mile for the county. When we consider that the number of persons per square mile for the whole of England and Wales is 618, and the number per square mile for the neighbouring county of Glamorganshire is 1383, it is evident that the county of Breconshire is very thinly populated. Indeed, with the exception of Merionethshire, Radnorshire, and Montgomeryshire it is the most sparsely populated county in Wales.

The diagram at the end of the book gives a graphic idea of the fluctuations in the population of the county since 1801. In that year the population was 32,325 and this increased in varying proportions to 61,474 in 1851. During the next ten years there was only an increase of 0·1 per cent. or 2 for every 1000 of the inhabitants. The next three decades were worse still, as they recorded decreases, and even in 1901 the increase was only 5 per cent. over the population in 1891.

How do we account for this? It is explained by the tendency of the young people of the villages to drift to the large towns in search of employment. There are no great industries in the county, so the people went outside its borders to find work. The fact that the old home industries of spinning, weaving, and the manufacture of domestic and industrial articles had been practically killed by the introduction of machine and factory-made goods, added to the rumours of fortunes that were to be acquired in the industrial districts, made emigration inevitable for those who could not find employment at home. Now however matters are becoming normal again. Emigration is lower, and the increase in the population follows naturally.

Many further interesting facts were shown in the 1901 returns, a few of which are given below. The people of the county lived in 12,775 houses and the males outnumbered the females by something over 1000. The people living in barracks numbered 268, nearly all of whom were males. In the workhouses were 217 inmates, whose comforts were seen to by 23 officials. In prison were 12 persons with nine officials to take care of them, and nine officials looked after the 13 persons in hospital.

The occupations of the people were also shown. Taking the males we find that agricultural pursuits were followed by the greatest number, but almost an equal number were employed in mining and quarrying. Next in number were the people engaged in building and works of construction, of whom carpenters reached the highest total. These were closely followed by men engaged in the conveyance of men, goods, and messages.

Those engaged in the following occupations numbered many less than those employed as above. Chief among them were the men employed in the manufacture of iron, the puddling and rolling of iron, steel smelting and founding, and in the manufacture of tinplates, while a few were engaged in combing, carding, spinning, and weaving wool. The manufacture of gunpowder, of chemicals, and of indiarubber gave employment to a small number of hands, and some were engaged in skinning and furriers' work, in tanning and in currying. Of course numbers were engaged in professional and commercial occupations. The females were mostly shown as domestic servants, dressmakers and milliners, and teachers.

The language spoken by the people was also shown in the returns. Taking the whole of the county, 54·0 per cent. spoke English only, 9·3 per cent. Welsh only, and 36·6 per cent. spoke English and Welsh. The remaining 0·1 per cent. of the population made no statement with reference to language in their census papers.

13. Agriculture — Main Cultivations, Woodland, Stock.

Agriculture, in former days, was in a very backward condition in Breconshire. Writers in the first half of the last century were loud in their complaints of the antiquated methods practised and the cumbersome implements used by the farmers of the county in their days. Needless to say matters are now much improved and, thanks to the advance of education and the efforts of Agricultural Societies, the farmers of Breconshire are as scientific in their methods as any, and use only the most modern implements on their farms.

We get our facts about the condition of agriculture in the county from the reports issued by the Board of Agriculture, and the figures quoted are mainly from the report issued for the year 1909. From a consideration of the report we arrive at three main facts which show us how little of the county is under cultivation. The first fact that strikes us is that 216,290 acres consist of mountain, heath, and waste land, which, though used as sheep walks during the summer, is in no other way productive from an agricultural point of view. The second broad division comprises some 200,000 odd acres which are devoted to crops and grass. Looked at casually we might infer that a considerable tract is under cultivation, but a more careful scrutiny reveals the fact that only about 39,000 acres of

Coed Farm, near Brecon

this land is arable, the remaining 161,000 acres being laid down as permanent pasture. The other broad division is the land devoted to the growth of wood, which amounts to 14,522 acres according to the 1895 returns.

Now that we have a general view of the state of agriculture in the county we can proceed to examine in detail the nature of the several crops. Cereals, or grain crops, are grown on 16,146 acres, or about 3½ per cent. of the land area. When we note that three-tenths or 30 per cent. of Essex is devoted to the same purpose we at once see how restricted the county is in this respect. Oats take up the greatest area, 10,580 acres being devoted to this purpose. Barley comes next with 3232 acres, and wheat third with 2282 acres. The other cereals grown are rye, beans, and peas, which take up 16, 21, and 15 acres respectively. Contrast the last two crops with the 47,323 acres devoted to them in Essex and the 61,000 acres in Lincolnshire and one is amazed that such a small area is devoted to their growth.

Green crops occupy 5822 acres. Of this area potatoes take up 794 acres, turnips and swedes 4573 acres, and mangolds 455 acres. Clover, sainfoin, and other grasses under rotation are grown on 16,192 acres, small fruit on 17 acres, bare fallow occupies 428 acres, and other crops 409 acres. The valleys of the county are admirably suitable for the growth of small fruit and increased attention has been paid to this cultivation within the last two years, so it is possible that later returns will show a welcome increase in the acreage devoted to it.

The county is well wooded, timber being at one time a considerable export, for Breconshire oak had an especially high reputation among shipbuilders. Timber is still felled and taken to the local saw-mills, but its consumption is practically confined to the district in which it is grown. Oak, elm, beech, sycamore, and ash are the most numerous of the deciduous trees, but there are also numbers of poplars and birches. A feature of the Usk valley is the frequent occurrence of the graceful Lombardy poplar. Willows, too, are numerous, sometimes attaining a large size. In the swampy parts alders abound. Large plantations of conifers are distributed over the county, larches and firs being the chief trees seen. There are also numerous Scots pines, especially in the east of the county, planted, so it is said, by Jacobites as a delicate expression of loyalty to the exiled Stuarts. Yews attain a large size

Cattle Fair, Devynock

and many are of considerable age, and there are also numbers of cedars.

The chief animal of the county is the sheep, 514,245 of which are reared on the pastures and on the mountains. Cows number 40,714, horses 12,825, and there are 7129 pigs. Good butter and cheese are made and large numbers of poultry are reared. These find a ready sale and good prices in the county markets.

14. Industries and Manufactures.

We have already read that the principal industry in the county is, and always has been, agriculture. Next in importance come mining and quarrying. Coal mining forms the most important branch and is carried on in the south-eastern and south-western corners of the county. In the south-east the coal mined is of the household kind, i.e. bituminous. This coal "binds" or cakes into a mass when ignited, and is known to the Welsh as *glo rhwym*. East of the Nedd, for a short distance, the coal is also bituminous, but less so than that in the south-east of the county, and is of a quality that makes admirable coke for use in blast furnaces. West of the Nedd the coal changes its character and contains a high percentage of carbon. It is the kind known as anthracite, and burns with intense heat and with little smoke and ash. The Welsh call it from its hardness, *glo carreg* or *glo caled*; it does not soil the fingers, and is peculiarly suited for the drying of malt or hops, and in places where heat is required without the disadvantages of noxious smoke and fumes. About 665,000 tons of coal, valued at £339,000, are raised annually.

The quarrying of limestone is carried on to a considerable extent, over 280,000 tons a year being obtained. A great deal of this is burned to make lime for industrial and agricultural purposes. The improvement of the soil by liming has been carried on for many centuries. Nearly 35,000 tons of sandstone are quarried for building purposes and also as flags for paving-stones, and setts for the curbing and channelling of roadways and pavements. A small quantity of slate is quarried, and 10,000 tons annually of clay, marl, brick-earth, and shale. From these latter bricks are made, and some fire-bricks are also manufactured from the fire-clay found in the south. Iron ore was formerly mined but of late years this industry has been discontinued.

Limestone Quarries near Vaynor

The working of iron, there is good reason to believe, was carried on in the county by the Romans. Proof of this is furnished by the masses of imperfectly fused scoriae found in different parts of the hundred of Crickhowell and known as Roman cinders. The modern revival of the industry took place over two hundred years ago, and with the development of the iron trade in South Wales several works were erected. These were established at different times in the Vale of Clydach in the parish of Llanelly, the Beaufort Works in the parish of Llangattock, Blaen Rhymney Works situated near the source of the river Rhymney, near Hirwain on the border of the county in the parish of Penderyn, and at Ynyscedwyn in the parish of Ystradgynlais.

Here the smelting and manufacture of iron was carried on to a considerable extent, especially on the border that adjoins Monmouthshire. The industry has declined of late years, and though some pig iron is made and some puddling and rolling of iron is carried on, the quantity of iron turned out is very small. There is also a little industry in the manufacture of tin-plates.

The fact that large quantities of wool are obtained each year from the sheep of the county has fostered the growth of a small woollen industry. The combing, carding, and spinning of wool into yarn, and the weaving of a coarse kind of woollen cloth, tweed, and flannel give employment to a number of the inhabitants. The industry is now chiefly carried on at Devynock, Brecon, near Llanwrtyd, and at Crickhowell. The mills and factories are small, and here and there weaving is carried on with small hand-looms in the homes of the weavers, who obtain the yarn from the mills. A leather of good quality is made in the county, especially at Brecon and Hay, and is used up locally in the manufacture of saddlery and in the shoe trade. Gunpowder and other explosives are made, the factories being situated along the banks of the river Mellte, and there are saw-mills at Talgarth and elsewhere.

The principal exports of the county are wool, butter, and cheese; of the former a quantity is still spun and knit into stockings in the hundred of Builth and in different parts of the highlands. There is also a considerable trade

Talgarth Saw-mills

at the county markets in poultry, eggs, etc., which are bought by hucksters for sale in the populous industrial districts. Sheep and cattle are sold in considerable numbers to dealers who attend the markets and fairs. Timber was once an important trade commodity but the trade is now practically extinct. Most of the country towns depend very largely upon the trade done with the surrounding agriculturists, but Brynmawr and Ystradgynlais are centres where coal mines or similar works provide a livelihood for the inhabitants.

15. Mines and Minerals.

We have already read of the most important minerals found in Breconshire, as the coal, the limestone, the sandstone and the brick-making earths, but we must take somewhat fuller notice of the coal measures, and there are a few other minerals to be mentioned, though they are not of very great importance.

The strata of coal and iron are the lowest found in the basin of the South Wales coalfield. This does not necessarily mean that the mines are of very great depth, as by some tremendous natural upheaval the edge of the basin containing the coal measures has been forced upwards so that it "out-crops," as it is called, in the southern portion of Breconshire. The upper measures have been worn away so that the lower ones are now comparatively easy of access. Coal occurs at three points along the border: (1) in the district contained by the Twrch, Tawe, Drum Mountain, and the Great Forest of Brecon; (2) in a corner of the territory from Blaen Rhymney to the northern side of Bryn Oer; and (3) in the district around the source of the Ebwy river.

The coal measures in the south-eastern corner have from 22 to 36 beds of coal, varying in thickness from

9 feet to 16 inches, interspersed among which are from 28 to 58 strata of iron-mine of varying thickness. The iron-mine strata in the thinnest parts have a collective thickness of 9 feet 6 inches, and in the thickest parts of 11 feet 9 inches. Three beds of fire-clay occur in this section.

In the south-western district the coal measures are more irregular, and the strata of ironstone vary from 1 inch to 5 inches in thickness, and frequently consist of irregular lumps called "balls of mine." The fire-clay beds in this district are of the best quality, especially those near the Dinas Rock in the Vale of Neath. For many years large quantities of the clay have been raised for use in the furnaces of the Neath and Swansea districts.

Breconshire contains no considerable quantities of any other ores. Sulphate of copper has been discovered on the northern confines of the county near the junction of the Elan with the Wye, and unsuccessful attempts were made years ago to discover a vein that might be worked with profit. Traces of lead ore have been seen in the south-west near the Dinas limestone, but no better results were forthcoming when search was made for it in workable quantities. Lead veins in the Llangion Hills, near Hay, were worked for a time, but as the cost of production left no margin for profit, the mines were abandoned. Small quantities of the ore have also been observed in other districts, but nowhere in workable quantities.

A mineral called Tripoli[1], or *Lapis curiosus*, was once worked in great quantities. It was found in the limestone

[1] Kieselguhr or Diatomite.

rock of the southern part of the county and was of a very pure quality. It was sent by the canals to Swansea, whence it was shipped to England to be used in the burnishing of metals. The trade in this commodity has now ceased. A black stone of close texture, called locally *Muchedd Irfon*, is found in the hundred of Builth. It was once used as a base or carriage on which the axles of engine wheels were supported, and was considered superior for this purpose to brass.

16. Spas.

The mineral wells of Breconshire have not attained the celebrity of the more fashionable Llandrindod Wells in the neighbouring county of Radnorshire, but they are fast becoming inland resorts that are much frequented during the summer season, by invalids who take the waters and by others for whom the attraction lies in the scenery, the sport, and the pure air which the neighbourhood of the wells affords. The wells all lie in the valley of the Irfon, at Builth, at Garth, at Llangammarch and Llanwrtyd. According to Sir R. Murchison the mineral nature of the waters is caused by the contact of eruptive rock with the schist, which produces much sulphuret of iron, the decomposition of which gives rise to their various mineral constituents.

The wells at Builth are at Park and Glanne, the former lying about a mile and a quarter to the north-west of the town, and the latter on the road to Cefn-y-bedd about a mile west of the town. The Park Wells are noteworthy for their saline water, which contains a high percentage of barium chloride. They are similar to the waters at Homburg and Kissengen though they lack that aeration characteristic of the waters of the continental spas. The presence of lithium, however, somewhat compensates for this want. There are also chalybeate and sulphur springs at Park. Glanne Wells are chiefly visited for the sulphur waters, though there is also a chalybeate spring. The Garth Wells are about six miles west of Builth and were only recently discovered. A magnesium well here is much advertised, and the waters are claimed to have beneficial results when taken for certain diseases.

Llangammarch Wells are also strongly impregnated with barium and other mineral constituents, and the waters are said to be similar to those of Kreuznach. Llangammarch lies eight miles west from Builth, and four miles further up the valley is Llanwrtyd, the best known and most popular of the Breconshire spas. Sulphur and chalybeate wells, with the fine air and scenery, attract large numbers of visitors during the brief summer season, but at other times Llanwrtyd is but a country village, devoid of any bustle, set among the wild hills of north Breconshire.

17. The Breconshire Reservoirs.

Some one has said that in Wales one can always hear the sound of running water, and to no county in the Principality is the statement more applicable than to Breconshire. Numerous streams of pure clear water flow from the hillsides in all directions, unpolluted by the refuse from any great industrial undertaking. It is no wonder, then, that many populous towns come to Breconshire for their water-supply and that so many important waterworks lie within, or are situated near, the confines of the county. Here, for their water, come Birmingham, Cardiff, Swansea, and Merthyr, while lesser works supply Neath, Aberdare, Ebbw Vale, and the larger towns of the county itself, and an immense scheme is under consideration whereby the Irfon valley will be tapped for the purpose of supplying London with Welsh water.

The most important of the waterworks are those constructed by the city of Birmingham in the valleys of the Elan and the Claerwen. Part, only, of these works lies within the county, the border-line between Breconshire and Radnorshire running along the line taken by the ancient courses of the streams in this part. The Caban Coch dam, formed at the junction of the Claerwen and the Elan, is the first of a series of dams which have transformed the picturesque Elan valley into the three huge artificial lakes of the Caban Coch compensation reservoir, and the Penygareg and Craig Goch

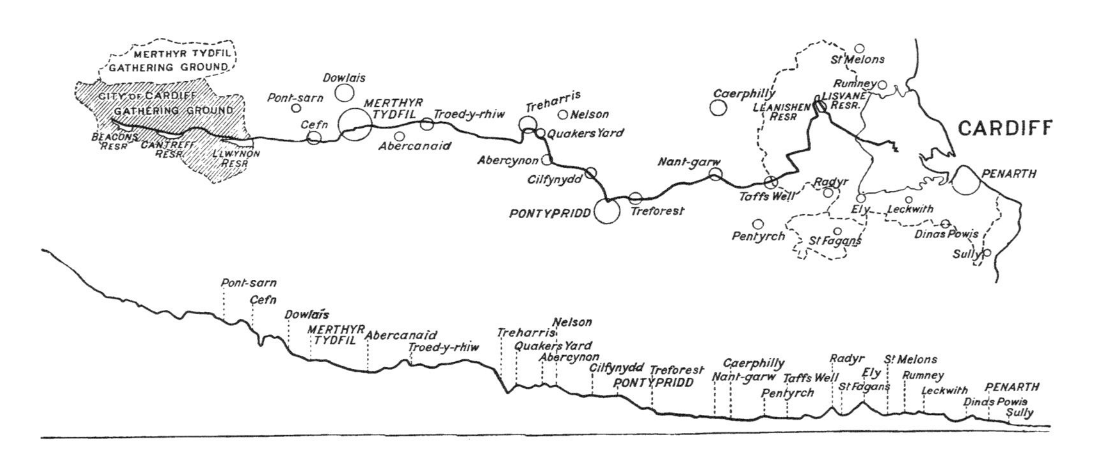

Plan and Section of Cardiff Water System

(By kind permission of C. H. Priestley, Esq., M.I.C.E.)

supply reservoirs. The first dam stands at an elevation of 820 feet, the Penygareg dam at an altitude of 945 feet, and the Craig Goch dam at 1040 feet, the water-surface area of the three reservoirs being 497 acres, 124 acres, and 217 acres respectively. Some 90 feet below the surface of the waters lie submerged the sites of Nant

Craig Goch Dam

Gwyllt and Cwm Elan houses, at both of which Shelley once resided, a church, a chapel and some twenty homesteads, all of which have been rebuilt elsewhere. At Caregddu, where the Caban Coch lake shoots out one arm to the north along the Elan valley and another to the south along the valley of the Claerwen, is a huge submerged dam from which the conduit leads. These lakes contain 11,000 million gallons of water, and yield a daily supply to Birmingham of 27 million gallons.

In the Claerwen valley is the Dolymynach dam, which impounds a mass of water with a surface area of 148 acres. Further dams are to be constructed in the Claerwen valley; one at Cil Oerwynt at an altitude of 1095 feet will create a reservoir of 269 acres surface area, and one at Nant y Beddau at 1175 feet altitude will impound a lake with 244 acres of surface. Altogether the city of Birmingham has acquired upwards of 40,000 acres of land in the watershed, thereby ensuring that the sources whence the city derives its water shall be absolutely uncontaminated. In all, this stupendous undertaking will cost some £6,000,000.

The aqueduct, from the intake to Birmingham, is some 73½ miles long. For the first 36½ miles it consists of a brick and concrete structure 9 feet in diameter, 13½ miles of which is in tunnel and 23 miles in cut and cover. For the remaining 37 miles the water is conveyed in huge iron pipes.

Up to 1851 Cardiff was dependent on water obtained from the canal, the river Taff, and from a few pumps in the town, one of the pumps being kept under lock and key during periods of drought. Under powers obtained in 1850, several reservoirs were constructed by a private company whose rights were acquired by the Corporation in 1879. Further local works were constructed, and in 1884 and 1894 powers were obtained to construct three large reservoirs in the Taff Fawr valley. Two of these reservoirs have been constructed and the third is now under process of construction. The Cantref reservoir, having a storage capacity of 323 million gallons, was completed in 1892, and immediately the Beacons reservoir, 35 miles from Cardiff, and having a storage capacity of 345 million gallons, was commenced, to be opened in 1897. These two reservoirs have surface areas of 44 acres and 52 acres respectively, and they furnish Cardiff with an annual supply of 2280 million gallons as well as 3 million gallons per day of compensation water into the Taff. Extended powers were obtained in 1909 for the reservoir now being constructed at Llwynon, which under the amended scheme will have a depth of 75 feet, a surface area of 144 acres, and a storage capacity of 1200 million gallons. The total cost of the Cardiff works, when completed, is estimated at £1,600,000, and they have been carried out under the direct supervision of the Corporation officials after two failures on the part of private contractors.

The Swansea waterworks in the Crai valley are curious, as they tap a river on the opposite side of the watershed from Swansea. This necessitated the construction of a tunnel 3 miles in length underneath a mountain terminating at Nantyrwydd. The dam across the valley is over a quarter of a mile in length, is about 100 feet in height at the centre, and has a width at the base of 80 feet. The works cost £800,000 to complete, and the reservoir has a capacity of over 1000 million gallons with a water surface area of over 159 acres.

The Merthyr waterworks are in the Taff Fechan valley and consist of two reservoirs. The upper, under the Beacons, has a capacity of 80 million gallons, and an older one, Pentwyn reservoir, 3 miles lower down, has a capacity of 350 million gallons. The Corporation are seeking further powers with a view of extending their works. The Nanthir reservoir, having a capacity of 40 million gallons and belonging to Aberdare, is also in the county; the Brynmawr reservoir on Cyrn Mountain has a capacity of 10 million gallons; Ebbw Vale is partly supplied with water from a reservoir in Llangynidr parish; and around the head waters of the Mellte reservoirs are in process of construction for the town of Neath. At Llwynfawr, Penderyn, too, works for supplying the Rhondda valley are in progress.

18. History of Breconshire.

Of the peoples who inhabited the district now termed Breconshire at the coming of the Romans and previous to that period, we have already read. The Silures, as the Romans called them, were not as civilised as the tribes to the eastward and so were harder to conquer, being more warlike in their nature and fuller of the love of freedom. They, like their brethren of Glamorgan, joined the forces of the valiant Caradoc and were among the defeated at the battle of Caer Caradoc.

The defeat and capture of Caradoc by Ostorius Scapula did not subdue these people, who, in their mountain fastnesses, defied the power of Rome until the coming of Julius Frontinus about the year 70 A.D. Even then, it is recorded, Frontinus subdued them as much by his diplomacy as by force of arms.

In accordance with their usual custom, the Romans set about the construction of their military roads and stations by means of which they were able to retain their hold on any subdued territory. We shall read of the remains of their works in a later chapter. For nearly four hundred years the Romans ruled this land, and when they evacuated it about 410 A.D., the native princes, who had been allowed to retain their rank while their conquerors ruled, entered into the government of their respective territories. We are told, that shortly after their departure, Brychan (400–450 A.D.) ruled in Garth Madryn, a district named after him Brycheiniog or Brecon. Brychan, according to the tradition, was the father of a numerous family, his sons and daughters being renowned for their piety and learning. Many of them became the tutelar saints of parochial churches, and thus imparted their names to the respective parishes. So revered was this family for its saintliness that it is mentioned in the British Triads as one of the three holy families of Britain.

The law of gavelkind obtained in those days and at Brychan's death his dominion was divided between two of his sons, Cledwyn and Rhian. The divided territory was again unified under Brychan's grandson, Caradoc Freichfras (Caradoc of the Brawny Arm). Caradoc, who lived towards the end of the fifth or the early part of the sixth century, was one of the Knights of the Round Table and fought under the renowned Arthur. His exploits at the famous battle of Cattraeth are preserved in the poem called the *Gododin*. His wife, Tegau Eurfron, also figures largely in the legends of Arthur's court. Caradoc was succeeded by his son, Cawrdaf, who in the Triads is named one of the three Prime Ministers of Britain.

Nothing is known of the history of Breconshire subsequent to the reign of Cawrdaf, until the reign of Teithwalch in the eighth century, which saw the first incursion into South Wales of the Saxons under the command of Ethelbald, King of Mercia. A battle was fought at Carno, near Rhymney, in the year 728. The Saxons came again in the reign of Tegyd, the son of Teithwalch, and, under Offa, they succeeded in wresting a considerable district consisting of the most fertile parts of Ferregs from the Welsh. Offa separated these wrested lands from the territories held by the Welsh by the huge earthwork called Offa's Dyke, that extended northwards from the river Wye in Herefordshire across the Marches.

Hwgan, or Huganus, was prince when Edward the Elder was King of England. Hwgan, calculating that Edward was fully occupied with the Danes, mustered an army and led them across the Saxon border. He was unexpectedly opposed by a powerful army under Ethelfleda, the Lady of Mercia, and was overthrown after an obstinate encounter. Ethelfleda followed up her victory by invading Hwgan's territory; she stormed his stronghold and carried off his wife and her attendants. Hwgan fled to the camp of the Danes and fell while assisting them against the Saxons. Hwgan's son, Dryfin, succeeded him, and during his time Athelstan wrested the whole of Ferregs from Brycheiniog and compelled Hwgan to pay tribute. About 944 A.D., when Dryfin was prince, a survey was made of the whole of Brycheiniog, in common with the rest of Wales, by the order of Howell Dda the King of Wales. Brycheiniog, in 982, was again invaded by a force under Alfred, Earl of Mercia, but after laying waste nearly the whole country, the Saxons were routed by the Welsh.

Roman Wall, Bannium

Dryfin's descendant, Bleddyn ap Maenarch, was the last native prince or lord of Brecheiniog. He was overthrown at the battle of Cwmgwernygad by a Norman host under Bernard Newmarch about the year 1091. Bleddyn was killed and his territories fell by right of conquest into the hands of Newmarch, who apportioned them among his followers, retaining a large portion for himself and the sovereignty of the whole. Newmarch, more just than the majority of Norman adventurers in Wales, granted the sons of Bleddyn portions of land for their sustenance and support, and treated the eldest son, Gwrgan, with much respect. To win the sympathy of the people Newmarch married a Welsh royal lady, Nest, the daughter of Gruffydd ap Llewelyn, Prince of North Wales.

The old Roman station of Caer Bannau (Bannium) was the capital of Brycheiniog, but this Bernard Newmarch demolished, using the materials of which it was built for the erection of his new castle of Brecon and of the town which grew up around its walls. Brecon was the seat of government for the Lordship Marcher of Brecon now founded, and from its castle the Lords Marcher ruled, with almost regal powers, over their Norman and Welsh vassals. Bernard Newmarch died sometime in the reign of Henry I, and was buried in Gloucester Cathedral.

The lordship and lands held by Bernard Newmarch became, after his death, the property of his daughter Sybil, who conveyed them by marriage to Milo Fitz-Walter, constable of Gloucester. FitzWalter was a firm supporter of the Empress Matilda and her son Henry, and as a reward for his services was afterwards created Earl of Hereford. Four of his sons succeeded him, but as they died without male heirs, their possessions went by marriage to Philip de Breos of Builth, the husband of their second sister.

The de Breos family held the lordship for many years, and some of its members figured largely in local and general history. William de Breos was the traitor who invited a number of Welsh chieftains to his castle of Abergavenny, and there murdered them. Other atrocious deeds he committed, but he and his wife were amply punished by the misfortunes that became their portion during their dealings with King John of England.

Reginald de Breos married Gwladys, daughter of Prince Llewelyn of North Wales, and assisted his father-in-law against King John. He was a treacherous ally and deserted Llewelyn to join his forces to those of the English king. The punishment meted him by the Welsh brought him to their side again, an action which King John naturally resented and punished him for by the deprivation of some of his estates. Reginald died in 1228 and was succeeded by his eldest son, William. William de Breos aided the English king in an expedition against Llewelyn but was captured by the Welsh prince, who released him on the payment of a heavy ransom. By some means, however, he incensed Llewelyn and, when he again fell into the hands of the Welsh prince, he was put to an ignominious death, and his lands wasted with fire and sword. Llewelyn made an attempt on the castle of Brecon, but failing to capture it, retreated after setting the town on fire.

Humphrey de Bohun now came into possession through his marriage to Eleanor, the second daughter of William de Breos. In 1265 the castles of Hay and Brecon were taken by Prince Edward, son of Henry III.

They had probably been in the hands of the Welsh, as at that time the territory round Builth was subject to Meredydd ap Rhys, to whom it had been given by Prince Llewelyn. Humphrey de Bohun was succeeded by his son Humphrey, the seventh Earl of Hereford. This earl's occupation of the lordship is noteworthy, as in his time a dispute arose concerning the exact boundary between the lordships of Brecon and Glamorgan, which developed into a local civil war. The vassals and tenants of the Earl of Gloucester and Lord of Glamorgan entered on the lands of the Earl of Hereford and carried away cattle and plunder. Reprisals on the part of the people of Brecon followed naturally, and some lives were lost. Matters were becoming somewhat serious when the king interfered and both nobles were fined and the liberties and privileges of the Lords Marcher were considerably curtailed.

During Humphrey de Bohun's life was enacted the tragedy that put an end, at least for a period, to the aspirations of the Welsh for their national independence. Llewelyn ap Gruffydd, the last native Prince of Wales, had been engaged in ravaging the territories of the supporters of the English king in Cardiganshire, and directed his course towards Builth through the valley of the Irfon. He crossed that stream at a bridge called Pontycoed and here he stationed a few troops from his meagre bodyguard to secure a retreat. At Builth he expected to meet with some of his allies and supporters, but soon discovered that he had been betrayed—by the very persons who had invited him to the district it is said—into the

Llewelyn's Monument, Cefn-y-Bedd

power of a force under the command of John Giffard and Sir Edmund Mortimer. Finding himself pursued by the English, he applied for succour to the garrison at Builth who, however, refused him aid. As he was returning to the place where he had left his followers he found that they had been attacked and overcome by the English, and before he could escape was himself discovered and slain by Adam de Francton, an English man-at-arms. The scene of his death was in a broom-covered dingle in Llanganten parish, and tradition has it that no broom has grown in that parish from that day to this.

The Humphrey de Bohun of Edward II's reign distinguished himself by his opposition to that weak king, and was killed at the battle of Boroughbridge in 1321. For a time his Welsh estates were held by the Despensers, but they again became the possessions of the de Bohuns only to be again vested in the Crown in the reign of Henry IV. During the rebellion under Owain Glyndwr, the peasantry were in sympathy with the Welsh patriot but the nobles were his bitter enemies. Owain burnt and destroyed the castle of Hay.

Owain Glyndwr, apart from the fact that they fought on opposite sides during the struggle for independence, had a personal quarrel with Sir David Gam of Castell Einion Sais. Gam, on one occasion, attempted to assassinate Glyndwr, who in revenge destroyed Gam's paternal residence of Castell Poytins and most probably also the castle of Einion Sais.

Glyndwr met with much success in Breconshire, so much so that Sir Thomas Berkeley was commanded to defend the county, with power to demand assistance from the sheriffs of six adjoining counties. The castles of the district were strongly fortified, and Henry IV himself must have been in Breconshire at one time, as a proclamation issued by him promising pardon to the rebels if they returned to their allegiance is dated from "Defynnock, September 15th, 1403."

In the reign of Henry VI, Henry, Earl of Buckingham, was Lord of Brecon, the estates of the lordship having descended to him from his mother, Anne, daughter of Humphrey de Bohun, to whom they had been regranted by the Crown. Buckingham ruled his lordship in a high-handed manner and was long remembered by the tenants for his tyranny and oppression. He was slain fighting on the Lancastrian side at the battle of Northampton in 1460. His grandson, Henry, when he attained his majority, was a strong supporter of Richard III, but Richard after winning the Crown forgot his obligations to the duke, who retired to Brecon determined on revenge.

Morton, bishop of Ely, a man who had a strong attachment for the murdered sons of Edward IV, had been committed into Buckingham's charge. The duke imprisoned him in the "Ely Tower" of Brecon Castle, but on reaching that place after the rupture with Richard III, Buckingham gave the bishop his liberty. More than this, he joined the bishop in a conspiracy to place Henry Tudor on the throne and marched a body of his retainers against the king. Henry Tudor was delayed by storms and the expedition failed, Buckingham being captured. He was afterwards executed at Salisbury in November, 1483. The estates were, of course, forfeited to the Crown, but the conspiracy did not fail in its effects, for two years afterwards Henry Tudor, accompanied by a small body of men, landed at Milford Haven and was met by Sir Rhys ap Thomas of Dynevor. Rhys, on a recruiting march, journeyed through Carmarthenshire and Breconshire, increasing his forces with every step. He met with such success that on his arrival at Brecon he was embarrassed with the numbers of his recruits and it became necessary to reduce his following. A selection was made; some were detailed for the advance, some for the defence of the lands left in the rear, and the remainder were dismissed to their homes with thanks for their offers of service. The battle of Bosworth Field followed, and a Welshman was crowned King of England on the field of battle.

Henry VII restored the Buckingham estates and honours to Edward, son of the late duke, whom he also made Constable of England. This Edward was executed for high treason in the reign of Henry VIII, and the lordship of Brecon with all its territories and revenues escheated to the Crown, in whose possession it remained for a considerable period.

During the reign of Henry VIII (1536) the present county was formed out of the Marches and it was enacted that in the whole of Wales law and justice should be administered in the same form as in England. The power of the Lords Marcher was thus curtailed, the almost regal jurisdiction they exercised was lost to them, and they were reduced to the condition of ordinary manorial lords. An active promoter of the petition for a more intimate union of Wales with England was Sir John Price, who, it is said, was the actual author of the petition presented to King Henry. Breconshire should be proud that Sir John lived at the Priory at Brecon, and was a native of the county.

Wales, on the whole, was on the side of the king during the Civil War, and Breconshire also was in the main Royalist. The county, however, was not the scene of much fighting, and the inhabitants, generally, managed to steer a course that kept them from figuring too prominently on either side during the struggle. Still, there was at least one battle, that at Brecon in 1645, where the Royalists lost some 700 men in killed, wounded, and prisoners.

King Charles I visited the county in August 1645, staying the night at the Priory, Brecon. Here he wrote the famous letter to his son in which he states the sad conclusion he had arrived at, that the young prince "must now prepare for the worst" and bidding him fly to France if he found himself in danger. We are told that the loyalty of the Breconshire people was not at such a low ebb as that of their neighbours of Glamorganshire; for they supplied sufficient horses to convert Sir Thomas Glenham's foot (which was under the command of Sir Henry Stradling of St Donat's) into dragoons. On the morning of the 7th of August, 1645, King Charles left Brecon for Radnor, dining on the way at Gwernyfed.

Though so much better than their neighbours, the

people of Breconshire must have been "shifty politicians," for in the same year of 1645 they petitioned Parliament to take them under its protection, eschewing royalty and promising obedience to Cromwell's commands. The leaven of loyalty was not wholly eradicated, however, for in May, 1648, they became violent Royalists again and arose, *en masse*, against the Parliamentarian, Colonel Horton, who had been sent into the county to put down some Cavaliers. The Restoration was hailed with joy by the people of the county.

19. Antiquities — Prehistoric, British, Roman, Saxon.

Perhaps the most interesting relic of prehistoric times is the site of the village of lake-dwellers discovered on the edge of the Llangorse Lake. This *crannoge*, as it is termed, is situated on the edge of a small island on the north side of the lake and was discovered by the Rev. E. N. Dumbleton in 1870. The discoveries include the piles on which was constructed the platform or pier supporting the huts, the piles indicating that they were placed in position by men of the Bronze Age. Numerous bones were found on the site and also some fragments of pottery and a ground stone. Traces of the hut-sites of prehistoric people have been noted in the neighbourhood of Builth.

Several stone circles in various stages of decay have been observed in the county. One stands on the high hill known as Y Gadair near Talgarth, and "Y Cerrig

Duon" stands on a hill to the west of Devynock. There are circles near Builth, and on Mannest, a hill near Llansantffraed, are the remains of another.

Of *cromlechau*, or dolmens, there were several, but those that remain are imperfect. One cromlech, a small

Remains or Cromlech near Crickhowell

one with capstone removed, stands in a field near Crickhowell, and the remains of Ty Illtyd, formerly consisting of three uprights bearing a sloping capstone, stands on Mannest. One cromlech remains near Llanwrtyd Wells and another on the Eppynt.

There are several British camps which, as a rule,

occupy the characteristic position of these strongholds—the brow or side of a hill. In Benni wood, near Brecon, is one almost entirely concealed by trees, with its ditch altogether obliterated. A large oval encampment crowns the summit of Pen y Crug, two miles north-west of Brecon,

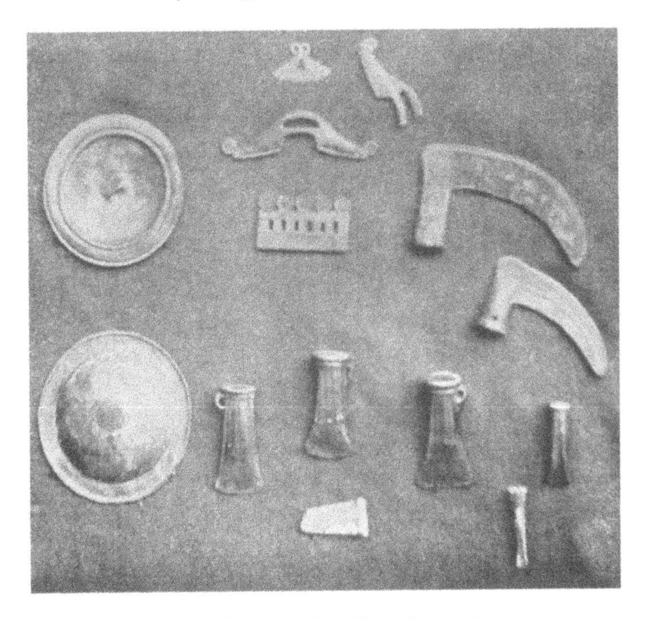

Ancient Bronzes found at Llwynfawr

and a triangular camp, defended by a ditch and ramparts of stones, stands in a position of great natural strength near Crickhowell. Other camps are situated at Alltfinio, near Glasbury, near Talgarth, and at Castell Dinas.

Remains of prehistoric days in the form of weapons

and implements are few. A ground stone, as just stated, was discovered in the *crannoge* in Llangorse Lake and a rudely formed spear-head of flint in a cairn near Bronllys. A coarse earthen vessel was also discovered in this cairn. A mound near Ystradfellte yielded a fine dagger-like knife of flint, a flint fabricator and several other flint chips, and a number of fragments of very rude pottery. The most interesting "find" of this type is the highly polished stone axe-hammer which was discovered on the hillside near Devynock—one of the largest axe-hammers discovered in this country. Some bronze implements were found in the Bishop's Meadow, near Ffynnonau, in 1882; and others in 1911, at Llwynfawr, Penderyn.

Carneddau, tumuli, or mounds are found in various parts, as on the scene of the conflict between the Saxons and Britons in 728; one, called Penmyarth, on a hill near Tretower, and others on the hills bordering Carmarthenshire. Mounds of artificial construction are also found near Dinas, Llanwrtyd Wells, Castell Madoc, Builth, at Ystradfellte, Vaynor, and Trecastle. Some of these have been opened, revealing *cists* or sepulchral stone chests.

The Romans had several stations in the county, the chief being that at Caer Bannau (*Bannium*), one of a chain of forts situated on the *Via Julia Montana*. Caer Bannau was constructed by Julius Agricola and was built to accommodate a garrison of one cohort with its complement of auxiliaries, or about 1000 men. Many interesting remains have been unearthed on the site, and on the line of the road, about a quarter of a mile from the fort, is the carved and inscribed stone known as *Maen y Morwynion* to be presently referred to. Y Gaer stands near Llanfihangel Cwmdu. It is somewhat smaller than Caer Bannau, and here also many relics of the Romans have been discovered. A third camp stands on Trecastle Mountain and it seems to have been a temporary station forming the base of operations for the conquest of Carmarthenshire.

Mound on Site of Roman Station near Crickhowell

A mound on the *Via Julia* near Crickhowell is said to be the site of a small Roman fort. Roman stones are preserved in many parts, though a number have disappeared. Several Roman altars have also been lost, and the Roman baths, discovered at Llanfrynach in 1783, were also not preserved.

About the year 1800 an interesting relic, in the form

of a quadrangular hand-bell of the type commonly used in the ancient Celtic church from the sixth to the ninth centuries, was discovered on a farm called Penydarren

Celtic Handbell

(*Welsh Museum, Cardiff*)

near the church of Llangenau. The bell was of sheet iron, riveted at the joints; it was provided with a handle at the top, and originally had been coated with brass by being dipped into the molten alloy. The tongue or clapper was missing. The bell disappeared, but when the Cardiff University College acquired the Salesbury library some years ago, a bell answering the description of the Llangenau bell passed into the possession of the Council of the College. This bell is now exhibited at the Welsh Museum, and authorities are of the opinion that it is the lost Llangenau bell.

Saxon remains are few. They consist chiefly of moated mounds on which the Saxon invaders erected their strongholds. On some of these the Normans erected their keeps at a later period in the history of the county. Where such mounds were utilised by the Normans, they will be noticed in the chapter dealing with the castles of the county.

20. Ancient Stones of Breconshire.

The ancient stones of Breconshire may be divided into the undecorated inscribed stones, and the decorated stones which may or may not have inscriptions cut upon them. Unfortunately several of these interesting relics have been lost, as the Valens stone at Tretower, the Vaynor stone, the crossed stone at Penymynydd, and the Roman stones known as the Cassanius stone and the Capel Coelbren stone.

Several of the Roman stones, however, remain. Undoubtedly the most interesting of these is that known as Maen y Morwynion, "The Maidens' Stone," which

Maen y Morwynion

is unique in Wales as it bears a carving in relief of a Roman soldier and his wife. The stone stands on the side of the road leading from Brecon to Aberyscir, and it had once an inscription which is now defaced with the exception of a few words and these even are very indistinct. The Roman stone at Battle is said to be one of the finest examples in the country. It is a fragment of a larger stone and has a well-cut inscription in Roman capitals. The stone was discovered in 1877 and experts date it from the end of the first or the beginning of the second century. At Tretower is the Peregrinus stone with the inscription "PEREGRINI FEC," and the Catacus stone at Llanfihangel Cwmdu has the inscription "CATACus hic jac(e)t filius Tegernacus" in mixed capitals and minuscules. Maen Madoc on the side of Sarn Helen has the inscription "DERVACI FILIUS IVSTI IC IACIT." The Turpillan stone, now at Glanusk Park whither it was removed after many vicissitudes, bears inscriptions in Latin and Ogam characters and dates from the third century. The Victorinus stone at Scethrog is cylindrical and is about three and a half feet long. When discovered it was in use as a garden roller and bears an inscription to the son of Victorinus in mixed capitals and minuscules. A highly interesting inscribed stone of ninth century date was discovered in Llanlleonfel churchyard in 1904.

The ornamented stones are extremely interesting as indicating the existence of a school of artists in stone at a very early period. These old British sculptors rendered intricate patterns with a high degree of artistic merit, the work on the Llanynys stone being exceptionally well done, but their depiction of figures is exceedingly crude.

The Llandefaelog stone is one of the few figured stones of the county and is interesting, as it was probably inspired by Maen y Morwynion. It stands in a wall in Llandefaelog churchyard and is divided into four compartments with ornamentations in relief. The first or uppermost compartment is occupied by a cross formed of two parallel raised bands interlaced at the junctions of the limbs and dilated at the ends of the limbs into trefoiled knots. In the spaces between the limbs are interlaced knots of ribbon-work. The second compartment contains a mutilated figure of a warrior, rudely represented in outline as wearing a long tunic and bearing arms. All available space is occupied by knots and patterns of ribbon-work. A rough cable pattern marks out the third compartment, which contained an incised inscription—now defaced—preceded by a cross. The bottom of the stone is entirely occupied by a double interlaced ribbon pattern of good design and execution. This stone is one of the rare instances in Wales where a figure of the deceased is represented on an early sepulchral slab.

Figures are also carved on the Llanhamlach stone. These are not of the deceased but are thought to represent the Virgin Mary and St John at the foot of the Cross, whose upright limb passes between the figures. Interlaced work and circular and straight-lined ornaments decorate the stone. The sides are also incised with ribbon-work and on the reverse face is an early inscription. Something similar to this stone is the one at Llanfrynach, where a human figure is carved with arms and fingers outstretched towards a small Greek cross on the upper part of the stone. Surrounding the legs of the figure and extending down the stone for about three feet is a ribbon-work pattern which terminates in another small Greek cross, two triquetra ornaments, and a small bird. The reverse side bears the name "iohir" in Anglo-Saxon minuscules. The Llangammarch stone has a raised cross of unequal arms contained within a circle, and among other decorations has a small incised figure with arms outstretched horizontally and a well-cut spiral. It is a fragment of a larger stone and is built into the west wall of the church.

The most graceful of the carved stones of the county, and one of the finest in Wales, is the Llanynys stone, now at Noyadd Sharmon three miles west of Builth. It is elaborately ornamented with interlaced ribbon-work of beautiful design and execution. Two sides only are exposed, the others being built into a wall. The upper portion of the face is formed into a cross with dilated ends to the limbs, while raised edges, divided into compartments or sections by knots, extend down the sides of the stone. The exposed side has a regular pattern carved upon it with the raised edge and knots also showing. The Llanddetty stone is ornamented with straight lines bent at right angles, the design looking like a Greek fret or a straight-lined maze. There is also an inscription in irregular and badly-cut letters.

A rudely-designed cross contained within an ill-cut circle is carved on the face of the Llangorse stone.

Ancient Carved Stones from Breconshire

(*From casts in Welsh Museum, Cardiff*)

There are rude ornamentations on each side of the stem of the cross and on the edge of the stone is a poorly-cut inscription, evidently two proper names. The encircled cross on the Trallwng stone shows much better workmanship, the lines being well cut and the circle accurately drawn. The edge shows Ogam characters forming a name which is also the first word of the Latin inscription cut upon the stone.

The Llanspyddid stone is known locally as the Brychan Brycheiniog Stone. The stone is of very early date and the church here is dedicated to St Cadog, a reputed son of Brychan, but there is no further evidence to support the tradition that here Brychan was buried. The stone is evidently a fragment of a larger stone and on it are two very simple crossed circles of unequal size, one above the other, the upper one being surrounded by four small circles. Some say the stone is a memorial to Awlach, father of Brychan.

The Llywell stone was discovered in 1878, ready to be set into position as a gate-post. The face of the stone is covered with incised markings of unique character, and the late Mr J. O. Westwood considered them unlike anything he had seen on Anglo-Saxon, Celtic, or Irish stones or manuscripts. On the upper part is cut a St Andrew's cross with circles, and in a corner is what may be the figure of a man or a bird. The ornaments and figures are much defaced and the Ogam characters on the edge of the stone are very indistinct. The ornament on the Devynock stone is curious, as it occurs at the end of the inscription instead of preceding it. It has a cross of equal limbs contained within a circle, with raised decorations between the circle and the end of the stone, suggesting a Maltese cross.

21. Architecture — (*a*) Ecclesiastical. Abbeys and Churches.

The religious houses of Breconshire, if we except the somewhat doubtful connection of Battle with the abbey of that name in England, were all situated in the county town. In the time of Henry I, Bernard de Newmarch founded a priory at Brecon for six Benedictine monks and made it subservient to the great abbey of Battle in Sussex. By him, his connections and successors, it was so liberally endowed that at the Dissolution of the Monasteries it was one of the richest of the religious houses of Wales. The lands of the priory, at the Dissolution, were granted to Sir John Price of Brecon and a portion of the buildings was by him converted into a dwelling-house. The priory church became the parish church, but the owners of the priory lands retained certain rights as to the chancel, which they still possess.

By the west gate of the town was also formerly a house of Black Friars, which after the Dissolution was converted by Henry VIII into a college, by the name of Christchurch, and connected with the College of Abergwili in Carmarthenshire. The college, now known as Christ College, was removed from Abergwili to Brecon and was at first ecclesiastical in character. It is now a public school, one of the two in Wales, and retains traces of its former ecclesiastical nature in its beautiful Early English chapel and a Divinity lectureship in the college. The college stands on the right bank of the Usk and comprises a fine group of buildings in the Early English style, some converted from their former uses and some modern. The college ranks high among the schools of

Brecon Priory Church

Wales and many of the distinguished sons of the Principality have been educated within its walls. (See p. 152.)

Breconshire contains many churches of interest, but the palm must be given to the priory church of St John the Evangelist, or the Holyrood at Brecon, which belonged to the Benedictine priory founded in 1095. The late

Professor E. A. Freeman considered this church to be indisputably the third church (not in a state of ruin) in the Principality, standing next in order after the Cathedral churches of St David's and Llandaff, and the choir furnishes one of the choicest examples of Early English architecture. Sir Gilbert Scott was entrusted with the restoration of this beautiful fabric, and the work was done in 1861 and between the years 1874–1875.

The church is a grand cruciform structure in the Norman, Early English, and Decorated styles of architecture. The eastern portions, including the chancel, transepts, and central tower, were built about 1220–1230. The nave and aisles were rebuilt or transformed during the Decorated period, but the piers of the nave arcade are of Norman date. Havard's Chapel, north of the choir, is also in the Decorated style. The church contains several monuments of great interest and some of the windows are also memorials to the dead.

The churches of Breconshire, outside the town of Brecon, are not remarkable for any striking architectural features. The country churches, with one or two exceptions, are severely simple in plan and perfectly plain in construction. They are built of native stone, chiefly sandstone, with, occasionally, dressings of imported stone where they have been restored of recent years. Yet the churches are full of interest, especially when we remember that the majority of the parochial edifices date their foundation from very early days.

The general arrangement of a country church of simple type is a western tower, a south porch, and a plain nave prolonged into a chancel with no architectural distinction to mark the division, save, perhaps, an ascent of one step. Llanfihangel Cwmdu church is curious, inasmuch as here we have an example with two south porches. The towers are often embattled, with

Bronllys Church
(*Showing detached East Tower*)

occasionally a stair turret prolonged into a kind of watch tower, their openings are mere slits, they are low, plain, and strongly built, and have altogether a semi-military appearance. Where the towers are not embattled the general arrangement is a low massive square tower hardly rising above the roof of the nave, with a top roofed over in dovecot fashion. A few of the towers have spires of the pyramidal or extinguisher type, and some of the churches have simple bell-cotes. Bronllys church is unique in the county, as it has a low, detached, dovecot-topped tower situated at the east side of the church. The

St Mary's Church, Brecon

naves are often long, narrow, and low, that of St Mary's church, Brecon, being exceptionally so. Sometimes an aisle is added, but to find two aisles is an exception.

Though the churches were founded at an early period there are no remains of any work before Norman times. In fact Norman churches are few, the Brecon churches and Bronllys being the only structures showing any great traces of Norman influences. The Early English period of construction is well represented, the majority of the churches owing the main portions of their structures to this period. The cruciform churches of the county belong to this period. They include the Priory church, Brecon, St David's, Llanthew—a smaller and plainer reproduction of the Priory church—and the parish church of Llanfihangel Abergwessin. Penpont, restored under the direction of Sir Gilbert Scott, is curious, as originally the west end was rounded, but the east end was square until the apse was built at the restoration of the church.

Crickhowell church is in the transitional Decorated style, and is built after the style of Llanthony Abbey, to which it was attached in pre-Reformation days. For a long period this church was the only one in the county possessing a spire, but pyramidal spires have been added to several during the last half century, when the majority of the buildings had undergone restoration. Llanelly church is also of the Decorated period. Perpendicular churches are few, though Perpendicular restorations as windows occur frequently, Brynmawr church being in this style, and there are Perpendicular windows in several churches. Perhaps the best examples of Perpendicular work are at Talgarth and Partishow, but even here some portions are older. Vaynor church has an interesting history. It was first built in the eighth century, to be replaced in 1287 by a second building having a saddle-back tower, after the destruction of the original structure during the War of Independence. It was a third time rebuilt in 1870.

Llangasty Talyllyn church, on the shores of Llangorse Lake, was the scene of the baptism of St Cynnog alluded to by Giraldus Cambrensis, but the ancient font is lost. Restored about 1849, the church is one of the prettiest in the county.

There are not many rood screens in the county,

Llangasty Talyllyn Church

numbers having been destroyed at one period or another, and probably some churches never possessed one. The finest example is in Patricio or Partishow church. This is a beautiful piece of work, equal to any in the country. There is also a good example at Llanfillo, and a very curious screen and loft, said to be of fourteenth century date, exists at Llanelieu. The screen alone remains at Llandefalle, Llanfihangel Nant Bran, Llanfihangel Cwmdu and Merthyr Cynnog. Most churches also lack a reredos. There is a mosaic and *opus sectile* example at Glasbury, a marble one at Llandefaelog (where the piscina and sedilia are also of marble and the chancel floor is of mosaic), and an oak reredos at Battle and Builth. Holy-water stoups remain in many churches, the best examples being at Aberyscir, Battle, Devynock, and Llanfihangel Bryn Pabuan, the stoup at the last-named church being surmounted by a quaintly-carved figure.

Many of the churches contain monuments to members of the families of the county. Battle church is supposed to have had some connection with Battle Abbey and this probably accounts for the number of monuments in so small a church. Several monuments are situated in the Priory church, Brecon; in Crickhowell and Glasbury churches; and at Builth there is an effigy in armour of sixteenth century date. There are also two brasses in Glasbury church. Llandefalle church has several slabs to the Vaughan family; Llansantffraed juxta Usk has a monument to Henry Vaughan the "Silurist"; and at Talgarth is a monument to Howell Harris of Trefecca, founder of Welsh Methodism.

Many of the churches have ancient fonts and some of them are of great interest. The font at Brecon Priory church is an elaborately decorated example, the decorations covering the bowl, stem, and lip of the font. The lip was also incised with an inscription which is now defaced and illegible. The Partishow font has an inscription which states that it was constructed in the time of Cynhyllyn who lived in the eleventh century. Semi-foliated designs are also cut on the font. Devynock font is ornamented with quatrefoils and oval pellets on the bowl, and trefoils at the angles of the base.

22. Architecture—(*b*) Military. Castles.

After the downfall and death of Bleddyn ap Maenarch at the battle of Cwmgwernygad in 1091, his conqueror Bernard Newmarch divided the subdued territory into fifteen manors, which he gave to the knights who had accompanied him on his expedition. The manors, with the names of their lords, were—Abercynrig and Slwch (Sir Reginald Awbrey); Crickhowell (Sir Humphrey Burghill); Tregunter (Sir Peter Gunter); Scethrog (Sir Miles de Picard); Llanhamlach and Llanfihangel Talyllyn (Sir John Walbeoff); Tredustan (Sir Humphrey Sollers); Pontwilliam (Sir Walter Howard); Trebois (Sir Richard de Bois); Peytin (Sir Richard Peytin); Bolgoed and Cray (Sir John Skull); Wernfawr (Sir Thomas Buller); Hay (Sir Phillip Walwyn); Aberyscir (Sir Hugh Surdwal); and Gilestone (Sir Giles Pierrepoint). Newmarch himself had the Lordship of Brecon and fixed his residence in the town of that name. Here he erected his castle with materials conveyed from the ruined Caer Bannau. His example in constructing a fortress was followed by all his knights and soon the conquered territory was dominated by a string of strongholds that made fast the grasp of the invader on the land. Several attempts were made by the Welsh to recover their lost lands but unsuccessfully, and they were compelled to submit to the Normans.

The distinctive feature of the Norman castle was the keep, a strong tower whose strength lay in its massive solidity. Not built for comfort, it was first and last a stronghold, ease being entirely subservient to military necessities. When surrounded by walls its situation was generally on the highest ground of the site, full advantage being taken of any natural defensive aid. The keep formed the last line of defence of the garrison if driven from the outer walls of the castle.

Norman keeps were of two types—the massive rectangular keep, and the lighter polygonal or circular shell keep. The interior of the rectangular keep was divided into three or four stories, in which were the rooms that formed the storehouse and quarters of the garrison. The entrance was generally on the first floor. The smaller shell keeps also contained apartments like the rectangular keeps, but the larger ones were merely open courts where shelters were erected for the lord and his garrison. The shell keep was often erected on the summit of an artificial mound, often the mound of a Saxon "burh."

During the period of castellation known as Early English the keep developed into the type known as donjon—a massive structure. Greater attention was paid to domestic comforts during this period, and the defences of the outer walls were more elaborate. Gatehouses and mural towers were provided, and long stretches of curtain walls and the bases of the round towers were defended by external wooden galleries. Then the central keep, surrounded by a ward enclosed within a curtain defended by mural towers and strong gateways, gradually developed into the concentric castles of the Edwardian period—the last word in medieval castellation.

The keep—the chief feature of Norman or Early English castles—was then dispensed with and its place was taken by an inner ward or court—a kind of glorified shell-keep—contained within a second and sometimes a third ring of defence, thus giving to this type of castle its concentric form. The chief features of the defences were strong gatehouses, massive curtain walls, and mural towers placed at the angles and at intervals along the curtains. From the towers every portion of the curtain could be swept by the fire of the garrison, thus doing away with the clumsy wooden galleries of the Early English period. Within the inner ward were the hall, chapel, kitchen, and the domestic apartments of the lord and his family, built against the curtain or in the towers and gatehouses defending the ward. Other apartments were distributed among the gatehouses and towers of the outer wards, and the whole castle was adapted for everyday use as a residence, as well as being a place of strength.

The ruins of many of these evidences of the Norman occupation still grace the scenery in many parts of the county. The Norman conquerors have passed away but the Welsh still remain on the soil, and strange to say most of the manors to-day are in the possession of gentlemen who pride themselves on a ancient Welsh descent.

Sketch Map showing the Chief Castles of Wales and the Border Counties

The accompanying sketch map shows how the castles were distributed in strategic chains, holding all the best and most fertile lands in the grasp of the Norman.

Blaenllyfni Castle stood at the head of the Llyfni river in such a situation as to command the important pass of the Bwlch, and it was probably built at an early date to check the raids of the restless Welshmen through the pass. The remains are few and are of post-Norman date, the castle, probably, having been rebuilt or extended about the reign of Henry III. It occupied a strong natural position, the swampy nature of the surrounding land adding to its strength. The castle was part of the possessions of the chief lord and is supposed to have occupied the site of the stronghold of Brec̓onmere, which was stormed by Ethelfleda, the Lady of the Mercians, after her defeat of Hwgan, Lord of Brycheiniog.

Brecon Castle was built about 1094 by Bernard Newmarch. The remains, which are those of the original castle, show that it was of considerable strength. The site was a strong one, an elevation on the western bank of the Honddu near its confluence with the Usk. There was here a Saxon mound on which the invaders no doubt had erected a stronghold during one of their incursions into the district. Most likely the existence of this mound influenced Newmarch when he decided to fix the main stronghold of his lordship at this spot rather than at Caer Bannau. The present remains consist of the keep and a portion of the curtain and a tower in the grounds of the Castle Hotel. The tower is known as the Ely Tower, and is so-called after the famous Bishop Morton of Ely, who was imprisoned in it. In 1404 the castle was garrisoned by 100 men-at-arms and 300 archers on horseback,

Brecon: Honddu Mill and Castle

under the joint command of Lord Audley and the Earl of Abergavenny.

Bronllys or Brynllys Castle stands on the left bank of the Llyfni, north of Talgarth, and its site commands the road leading from Brecon to Hereford. The present remains consist of the keep of the original castle which was built on a large mound, the situation of a former fortress. The tower is cylindrical in form, and consisted of three floors and a vaulted basement. The top was battlemented, and was probably supplied with fittings for wooden galleries placed externally to defend the base of the tower.

The origin of the tower has given rise to many fanciful theories. Some believed it to be British and contemporary with the round towers of Ireland; but the general structure of the keep indicates that it was built in the Early English period, about the first quarter of the fourteenth century, with alterations and repairs of Decorated and Perpendicular date. The castle has had various owners, among whom were the de Bohuns, the Staffords, and the Crown. About 1450 it was the home of Bedo Bruinllys, a well-esteemed bard of that period. It is now in the hands of a private owner, and a modern house occupies a portion of the site, incorporating some of the remains of the castle within its walls.

Builth Castle was originally constructed in the ninth or early in the tenth century, and, as Mr G. T. Clarke suggests, was probably a Saxon outpost thrown up either during the wars that preceded the construction of Offa's Dyke or to avoid the aggressions which followed it, the plan of construction leading rather to the latter conclusion. When Bernard Newmarch came he occupied the mound,

Bronllys Castle

and either he or one of his immediate successors, more probably one of the latter, constructed upon it defences of the late Norman or Early English type. The castle occupied a dominating position above the town and commanded the passage of the river over the old bridge. Few

Earthworks, Builth Castle

traces of the masonry of the stronghold now remain, but we see that its form was nearly a circle and that the castle consisted of a keep and an outlying ward contained within curtains defended by towers.

Like the castles of Brecon and Builth, Crickhowell Castle was originally a Saxon moated mound with a ward at its base. It stood between the town of Crickhowell and the Usk. Its remains, which are not very considerable and not of any great interest, consist of two towers

Crickhowell Castle

of the Edwardian period, one a drum or round tower, and the other a rectangular tower. They are situated on the curtain which the Norman erected around the ward at the foot of the mound. Chief of the other masonry are the ruins of the gatehouse that defended the entrance of the ascent into the keep.

The remains of Dinas Castle stand on the summit of a high hill commanding the pass between Talgarth and Crickhowell. Who erected it is unknown and nothing is certain of its history. Mr G. T. Clarke described it as a hill castle, in form much resembling Morlais in Glamorgan. Its position also resembles that of Morlais, as it commanded an outlet into the valley of the Usk as Morlais commanded an outlet into the Taff valley.

There are no remains of the castle of Einion Sais and in fact its actual site is unknown. It is only known that it was situated near Capel y Bettws, some 1½ miles west of Brecon. The castle took its name from Einion ap Rees, called "Y Sais," the Englishman, because he took service with the King of England and resided for many years in that country. Under Edward III he was present at the battle of Crecy in 1346, and he also fought at the battle of Poictiers in 1356. Sir David Gam, the great Breconshire hero of Agincourt[1], was his descendant, and as Sir David was a strong and bitter opponent of Owain Glyndwr, it is practically certain that Einion Sais Castle was totally demolished by the Welsh prince.

Hay Castle stood in the town of Hay, near the borders of three counties. The town was also walled and had three gates, and both castle and town figured prominently

[1] "Edward the Duke of York, the Earl of Suffolk,
Sir Richard Ketley, *Davy Gam Esquire*:
None else of name; and of all other men
But five and twenty." *Henry V.* Act IV. 8.

in matters political and military previous to the death of Prince Llewelyn in 1282. The remains of the castle consist of a large square tower, probably the keep, and a gateway that adjoins the tower to the east. The gateway, still in a fair state of preservation, is a Gothic structure, and had the usual provisions for a portcullis and other defences of a gate. It was situated outside the town, and as it was thus beyond the jurisdiction of the ancient baron court, it formed a kind of sanctuary for debtors who fled for temporary refuge from creditors and officers of the law. The manor of Hay, at first a possession of the Walwyn family, soon passed into the hands of the chief lord, and went with other estates appertaining to the lordship until the reign of Henry VIII. Archbishop Baldwin, when preaching the Crusade in 1188, spent a night in Hay Castle. He was accompanied by Gerald the Welshman, archdeacon of Brecon. The castle was frequently the centre of strife and suffered considerably at the hands of contending forces. Henry II is credited with its destruction during one of his expeditions into Wales, and towards the end of his reign it was rebuilt by William de Breos. De Breos's wife was Maud Ste Valerie, and this lady, according to the local tradition, performed the superhuman feat of rebuilding Hay Castle in a single night, carrying the stones necessary for her purpose in her apron. Maud, also known as Maud de Haie, had the sobriquet of Moll Walbee, and her alleged feats gave rise to many parochial legends in several parts of the county. Time after time Hay Castle was destroyed by English or Welsh, only to be again rebuilt, but after its destruction by Owain Glyndwr it does not seem to have been again reconstructed, though the Glanusk dower house occupies a portion of the site.

A mansion, erected in 1588 by Thomas ap Howel, is known as Castell Madoc, but the site was formerly occupied by an older fortified place. Its situation is on the Honddu above Brecon, and Mr G. T. Clarke says it was, more probably, an ancient British fortress. The mound of the ancient keep remains within the grounds. Scarcely any vestige remains of Pencelli, or Penkelly Castle, which stood four miles south-east of Brecon. The materials of this ancient structure were used by the Herberts, in the fifteenth century, for the erection of a castellated mansion, now in ruins and partly converted into a farmhouse.

Castell Du, Rhyd y Briw, or Rhyd y Briew Castle stood upon a small knoll upon the western bank of the river Senni near its fall into the Usk and a short distance from Devynock. It is said that it was erected in the reign of Edward III to protect travellers and the inhabitants of the surrounding district from the ravages of the outlaws of the hills and forests. The castle seems to have consisted only of a tower surrounded by a walled court and was a place "where the robbers from the mountains were confined and frequently executed without trial." Scethrog Castle, or tower, stood a mile north-east of Pencelli Castle, and portions of its walls are incorporated into those of a farmhouse which has been erected upon the site. A modern mansion of the same name stands a short distance from the site of the ancient fortalice.

The tower of Talgarth has been considered by some

writers as merely a structure intended for use as the old borough gaol. The tower, however, was really a strongly fortified structure, similar to the "peels" of the Scottish border, and when it was no longer required for strictly defensive purposes was used as the borough prison. It stands on the banks of the Llyfni and commands the passage of the river at this point, and also the pass leading to Crickhowell. The tower was built in the fourteenth century, and its defences were strengthened by the provision of machicolations, part of which remain. No traces are visible of any of the usual external defences of a castle, and the tower appears to have been entirely isolated. A military building of this type, though once fairly common on the borders of England and Scotland, is a rare occurrence in Wales.

Leland speaks of the tower as standing in "Englisch Talgarth," which reminds us that many of the manors held by the Normans were divided into two portions, inhabited respectively by Welsh and Norman tenants. The former held by *Welsherie*, or the ancient Welsh tenure, and the latter by *Englisherie* or English tenure, according to the strict rules of the feudal system. The Welsh often won this great privilege, for a great privilege it was, by rebellion, and their lot was much easier than if they held their lands from their lords according to the feudal tenure.

Tretower Castle was built on the left bank of the river Rhiangoll, about a mile and a half above its junction with the Usk. It forms one of a chain of castles (Blaenllyfni, Dinas, Crickhowell, and Abergavenny

Tretower Castle: the Keep

in Monmouthshire were the others) that checked the Welsh of Breconshire and Radnorshire from advancing to the south, and made their expeditions in that direction hazardous adventures. The site of the castle is, naturally, a strong one, being on a gravelly eminence covered on three sides by a swamp, so that approach was difficult except from the north-east. The castle ruins exhibit many architectural features of great interest. One probably unique feature is the fact that an Early English tower of cylindrical formation has been erected in the centre of the square enclosure of a previously-built Norman keep. Mr Clarke comments on it thus: "Tretower is a rare, probably a solitary example, of a rectangular Norman keep, which has been gutted, and its central part occupied by an Early English tower." The final destruction of the castle was the work of Owain Glyndwr.

23. Architecture—(*c*) Domestic.

Castle-building reached its highest point of development in the reign of Edward I, but by the time of Edward III the need for such strong fortresses had passed away. Two factors brought this about, the more settled state of the country, and the introduction of gunpowder, which made the reduction of a castle a far easier matter than in older times. Still, some defence was necessary against the bands of robbers unprovided with cannon, so a compromise was effected in the fortified manor houses, which gave sufficient protection against sudden attacks by small bands and yet allowed room for the comfort and luxury that an advanced state of civilisation demanded.

The close of the Wars of the Roses, which saw, with the accession of the Tudors, still greater security to life and property, brought about a more luxurious and elaborate style of domestic architecture, and the manor houses of this period were buildings of great beauty and much comfort. The earlier manors of this period were built in the form of a quadrangle with the hall in the middle and wings on the sides. In the reign of Elizabeth, the ground plan was often like the capital letter "E," out of compliment to the virgin Queen. Beautiful chimney pieces, tall chimneys, wide staircases, ornamented plaster work, and high oak wainscotting, with the characteristic windows and doorways of the Perpendicular style of architecture were features of the mansions of this date. As there was plenty of stone of various kinds in Breconshire it was used in the construction of houses, and "tilestones" of thin slabs of sandstone or shale were in general use for roofing purposes.

In earlier days numerous manor houses of the Tudor period were scattered through the fertile valleys of the county, inhabited for the most part by descendants of the ancient owners of the soil. During the last century or so a great change has taken place. Much of the land has passed from the hands of the ancient families into the possession of others, and many of the ancient houses have either disappeared, or have been allowed to fall into ruins, or have been reduced to the state of an ordinary farmhouse. Still, many of these ancient edifices remain, though modernised to suit the luxurious mode of living of a softer generation. In the valleys and in the suburbs of the few towns modern residences have sprung up, which, though unconnected with the traditions of the past, are, one must confess, more convenient and comfortable than the structures of bygone days.

The fate of Tretower is typical of many of the ancient houses of the county. Here a quadrangular manor house, said to be of the time of Edward III, fortified by a strong gateway and stout walls, has been allowed to fall into decay. The courtyard is now the rubbish-heap of a farm, the buildings on the right as we enter, with the remains of a once graceful timber balcony, are the stables and outhouses, while the hall, yet retaining much of its beautiful woodwork and timber roof, has become the receptacle of hay. The remaining side is occupied by a later farmhouse, which, though not so incongruous as some others may have been, yet rises in condemnation of the people who allowed the fair place of Henry Vaughan to become an appendage when it should be the chief feature of the site.

Newton is perhaps the most interesting specimen of the buildings of Elizabethan times. It and Abercamlais are typical examples of the late Tudor manor house. Newton was erected in 1582 by Sir John Games, and is a strong and not unsightly mansion that is half fortress and half domestic residence. Trebarried, built over two hundred years ago by William ap Henry Vaughan ap Fychan, lies in a sheltered position and is now in the occupation of a farmer. For many years it contained several interesting portraits of the old family of Vaughans who inhabited it. Treberfedd, in the neighbourhood of Llangorse Lake, is a large and imposing building of Elizabethan date, which has been considerably renovated and enlarged. Gwernyfed, also of Tudor foundation, was visited by Charles I, who also stayed at Brecon Priory, another old Tudor house.

Treberfedd

No traces remain of the castellated mansion of the Herberts at Crickhowell except an old gateway in the decorated Gothic style of architecture which formed the entrance into the quadrangle. A modern house now occupies the site of the old mansion, but unfortunately it is entirely out of harmony with the beautiful old gateway.

Aberclydach is a very ancient mansion. It was occupied by the descendants of Rhys Goch, and was subsequently the home of the Lewises of Aberclydach, a well-known Breconshire family.

Porthmawr Gate, Crickhowell

The old Castle Madoc as we have seen was a true castle, and the mound of the keep is still visible in the grounds. The present mansion bearing the name was originally erected in 1588 by Thomas ap Howel, whose descendants occupied it until the year 1796. The last male representative of the Powel or ap Howel family, Charles Powel, died in that year, and on the death of his daughter the mansion and estates passed into the hands of a cousin, named Hugh Price, from whom the present owners are descended.

Aberyscir, situated close to the confluence of the Yscir with the Usk and near the ancient fortress of "Y Gaer," was erected about 1571. The mansion of Abercamlais was also erected about 1571, but has from time to time been considerably altered and enlarged. Abercamlais was the old home of the well-known Williams family. A branch of this family lived at Penpont, a mansion situated in the Vale of Usk above Brecon. The grounds contain some fine specimens of cedar and fir, and in the neighbourhood is the site of the castle of Einion Sais, of which no trace now remains.

Craig y Nos, the residence of Baroness Cederström (Madame Patti), stands in a picturesque situation on the banks of the Tawe in the parish of Ystradgynlais. It is a large modern mansion of stone in the Italian style with a private theatre attached which is capable of seating an audience of 220 persons. In the same parish is the ancient house of Ynyscedwyn.

The farmhouses and cottages of the county are generally well built, are two storied in height, and are constructed of local stone—Pennant Sandstone or Carboniferous Limestone in the Carboniferous district, Old Red Sandstone in the Devonian region, and shale and sandy flagstones on the Silurian formation. The softness of the stone in the last district gives the houses a weather-beaten appearance, and plastering and other methods are adopted to prevent the decay of the stones. The use of whitewash, or of a red or yellow ochre, is prevalent everywhere; most of the country cottages and farms being coated with a thick layer of one of these substances.

Cottages on Silurian Rock

Where the houses have been repaired of late years slate roofs have superseded the shale slab and sandy tilestone roofs used in older days, but many of the houses retain their original roofing and it is by no means uncommon to see the front portion of the roofs covered with slates while the back slope retains the stone slabs. Square dormer windows are a common feature of the older houses in the southern part of the county and they also occur, though more rarely, in the north. Thatched roofs are very rare, only a few instances being met with, and these seem to be confined to the extreme north of the county. The farmbuildings are generally set in a quadrangle with the house facing the south, or the south-west, and the courtyard in front. The outbuildings are, as a rule, well-built stone structures, almost invariably retaining stone slab roofs, but stone tiled buildings with sides of long slabs of timber set upon a foundation of masonry are to be seen in the north of the county. Brick buildings are rare in the country districts and are not by any means common even in the towns.

24. Communications—Past and Present. Roads, Railways, and Canals.

The earliest means of communication in Breconshire were undoubtedly the trackways worn in the valleys and on mountain sides by the feet of the ancient inhabitants of the county, but no trace of these remains, though perhaps we may guess the general directions of some of them by those of later roads and byways. Formerly every valley or dale had its trackway or rough bridle road, but in early times, and especially when the country was largely covered with forest, travellers from one settlement to another usually kept higher up on the sides of the hills.

The first made roads of the county were those constructed by those excellent road-makers, the Romans. The making of these causeways was an important factor

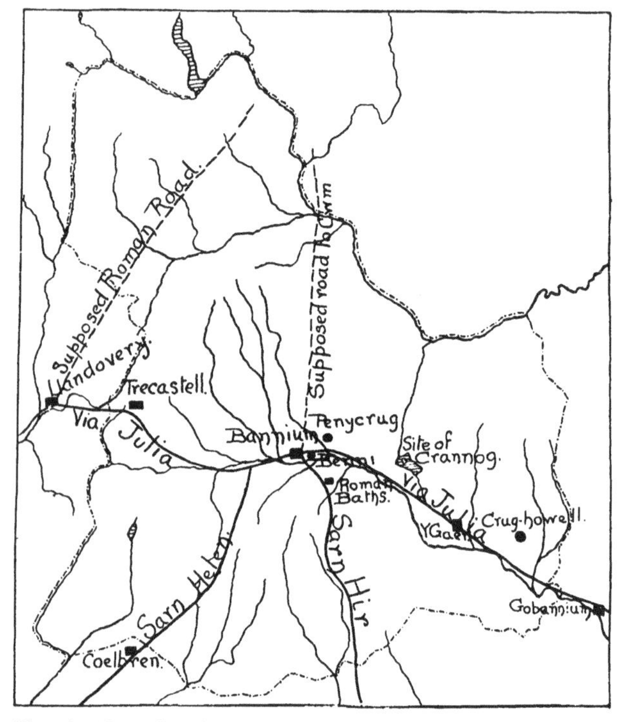

Sketch Map showing British Camps and Roman Stations and Roads

in their tactical scheme for keeping the country under subjection, for stimulating commerce, and for facilitating the collection of revenue. They ran from one important station to another, were defended in places by minor stations, and were marked by mounds and mile-stones.

The principal Roman road in Breconshire was a branch of the *Via Julia Maritima*, a road made by Julius Frontinus along the southern coast of Wales. The branch road through Breconshire was called the *Via Julia Montana* to distinguish it from the main road from which it diverged. The *Via Julia Montana* branched from the other at *Isca Silurum*, or *Legionem*, the station of the Romans at the place now Caerleon, and passing *Gobannium* (Abergavenny) entered the county south of Crickhowell. Passing Crickhowell it went through Tretower to a station, now called "Y Gaer," situated on an eminence near Llanfihangel Cwmdu. From Y Gaer it went over a pass called Y Bwlch, and following a line a little northward of the present main road between Crickhowell and Brecon, arrived at the latter town and passed on to the Roman station of *Bannium* (Caer Bannau). Beyond Caer Bannau it crossed the Usk, and following that river valley westwards, proceeded *via* Rhydybriw into Carmarthenshire, connecting with the *Via Julia Maritima* at *Maridunum* (Carmarthen).

Several cross roads connected the great central station of Caer Bannau with other stations in various parts of the country. Sarn Hir was one of these. It ran northward from *Tibia Amnis* (Cardiff) and entered Breconshire at Brynoer, and continuing in a straight line, crossed the Usk, and joined the *Via Julia Montana* at, or near, Caer Bannau. The Sarn Helen, or *Via Helena*, in the same manner connected Caer Bannau with *Nidum* (Neath). This road entered the county at Ton y Fildra and proceeded across a brook called Nant Hir to Blaen Nedd. It pursued a course parallel with the road from Pont Nedd Fechan to Brecon for about a mile and passed the stone called Y Maen Llia. From this spot it descended the hill on the southern side of the Senni river. Its further course is lost until near Blaengwrthid, where it is traced for a short distance only to be lost again. It is conjectured that it joined the *Via Julia* near Penpont, in the Vale of Usk. A road, sometimes called *Via Devana*, is supposed to have connected Caer Bannau with the station at Cwm in the valley of the Ithon in Radnorshire and to have proceeded thence to the station of *Deva* at Chester. No traces of it have been observed in Breconshire. Some traces of a Roman road have been observed in the northern part of the county: it is supposed to have been a crossway running between *Maridunum* and the station at Cwm.

After the departure of the Romans the roads fell into disuse and consequently into disrepair, and no doubt their destruction was accelerated by the transport of their materials for other purposes. Vegetation overgrew the paved ways, and they now can only be followed in part. For hundreds of years the roads of the county were mere farmers' lanes and bridle paths, but towards the end of the eighteenth and the beginning of the nineteenth century a general improvement set in. New roads were made and the existing tracks that were important enough were placed in a serviceable condition. Now good roads branch off from the county towns in all directions, and only in the most remote districts is locomotion difficult.

Perhaps the most important road still is the old coach road from London to Milford. It enters the county south of Crickhowell and runs in a north-westerly direction to Brecon. From the county town it proceeds westwards through the Vale of Usk, and enters Carmarthenshire *via* Trecastell. A branch of the road from London to Radnor runs from Hereford, and passing through the town of Hay, proceeds to Brecon. A road connects this branch with Builth through the Wye Valley, while Builth is also connected with the county town by a road through the valley of the Honddu. Another good road runs through the valley of the Irfon from Builth to Llandovery in Carmarthenshire. These are the main roads, but there are numerous cross roads running in all directions. The chief roads without exception run mainly along the river valleys, and when they do cross a mountain range they only do so to drop into other valleys on the opposite slopes. The peculiar river system with the numerous branching tributaries made it obvious where and how the modern roads should run.

Railways have long superseded canals, but before the advent of steam-traffic considerable trade was done over these artificial waterways. The Breconshire Canal was opened for traffic in 1801, and at first ran from Brecon to Clydach, but in 1811 an extension to Pontypool was constructed. At Pontypool the canal joined the

Monmouthshire Canal, thus affording facilities for transport from Brecon to the Bristol Channel. The Swansea Canal, running through the Tawe Valley, has its northern terminus at Hen Neuadd, in the parish of Ystradfellte, about four miles of the waterway lying in the county.

The chief means of communication nowadays are, of course, the railways. Tramways, having horses as the motive power, had been in use in the coal and iron

The Canal at Brecon

districts before the introduction of the locomotive. From the beginning of the reign of the late Queen Victoria to the present time the construction and use of railways have developed so much that now the county has a good railway system connecting it with the other counties of Wales and with England. Like the roads, the railway lines, with one exception, radiate from the county town.

The Cambrian Railway has a branch from Moat Lane in Montgomeryshire into the county. From Moat Lane it runs southwards to Rhayader in Radnorshire, and then through the valley of the Wye, the Radnor-Brecon boundary, to Builth Road, to Three Cocks, to Talgarth, and thence to Talyllyn and Brecon. The same route, from Three Cocks through Talyllyn to Brecon, is also followed by the Midland Company's Hay and Brecon branch, which, continuing over the Neath and Brecon line, runs through Devynock and Coelbren to Neath and Swansea. The Brecon and Merthyr Railway runs from Brecon to Talyllyn, and thence to Talybont. It follows the Collwng and Taff Fechan valleys to Pontsticill Junction. Here a short line branches off to Merthyr, and the main line proceeds, *via* Dowlais, through the Rhymney valley to Bargoed and Newport. The Central and South Wales branch of the London and North Western Railway enters from Radnorshire at Builth Road, and proceeds down the Irfon valley, through Llangammarch and Llanwrtyd into the Towy valley, where one branch turns off to Swansea and another continues down the Towy valley to Carmarthen.

25. Administration and Divisions—Past and Present.

The ancient form of government in Wales differed from that of England insomuch that it was tribal, though the tribes were subject, often only nominally so, to an overlord. The Principality was divided into a number of small kingdoms or principalities of greater or lesser extent which were grouped under the sovereignty of the overlords, who were at first the kings of Wales, and afterwards the kings or princes of the three great sub-divisions of Wales after the death of Rhodri Mawr. In the division made by Rhodri, Breconshire was included in the kingdom of Deheubarth, or South Wales.

For purposes of administration these districts were divided into cantrefs and commots. The cantref was somewhat akin to the English hundred and was composed of two or more commots; the commot was the unit of government. It was thus akin to the English manor and was governed by a lord and his officers—the *maer* or steward, and the *canghellor* or chancellor. Socially, the commot was composed of the tribesmen—who were of the kin of the lord, held the land by gavelkind and were distributed in homesteads on the land held by their "families"—and the "strangers," who lived in villages or communities under the control and supervision of a representative of the lord.

The "strangers" were partly composed of the pre-Cymric inhabitants of the land, and were not allowed the privileges of holding land, of bearing arms, or of hunting. To them was allotted the task of cultivating their lord's land and performing certain menial duties. Below the "strangers," or non-tribesmen, were the slaves.

It must be pointed out that the tribesman did not hold possession of the land as an individual but as a member of a family, and herein lies the great difference between the English system and the Welsh. The tribesman could claim pasturage, a share of the stock and of the farm produce, but before he could do so he had to perform his share of the labour. Strictly speaking, all he had was the use of the land. Every now and then, according to strict rule, the land of a "family" was redistributed, and new "families" were formed, but this is a lengthy subject which we have no space to deal with here.

After the Norman settlement many attempts were made to introduce the feudal system, but among the purely Welsh element not with any degree of success. The manorial system obtained when the Normans came into possession, but the manors were never grouped into a shire, and each lord of the manor ruled almost as he thought fit. In many manors the Norman tenants held under the feudal system, while the Welsh tenants held under their ancient tenure. We have already given an example of this in the Manor of Talgarth, where the manor was divided into Welsh Talgarth and English Talgarth, and separate courts (*Welsherie* and *Englisherie*) dealt justice to the two races.

Some portions of Wales were made shire ground by Edward I, but Brycheiniog was under the rule of the Lords Marcher until the reign of Henry VIII. The formation of Breconshire meant more than the defining of the boundaries of a certain tract of land and the christening of that land with a name; it meant the institution of a certain form of government. The officers of justice were now the representatives of the King and not of the lord as was the case when the Lords Marcher ruled.

A breach of the peace was a Breach of the King's Peace, and his officers punished the offenders who committed that breach.

The chief officials of the new shire were the sheriff, coroners, and bailiffs. Each commot had its own coroner and bailiff, and the sheriff paid periodical visits to the commot to try minor offences. Serious offences he sent to the county town for trial by the King's Justices, who attended periodically for that purpose.

The formation of the Marches into shire ground and the incorporation of the Principality with England did not do away with all distinction between the English shires and the Welsh shires. The latter were still under some disadvantages. An attempt to better the condition of Wales was made in 1545, when the Great Sessions of Wales were instituted. These formed a High Court of Justice for Wales, independent of the Courts at Westminster, and in 1576 the Council of Wales was instituted. The Great Sessions and the Council were abolished in 1689, bringing the Administration of Wales into line with that of England, and the final difference was abolished in the first decade of Queen Victoria's reign, when the method of selecting sheriffs for the Welsh counties was made similar to that of England.

The chief officers of the county are the Lord Lieutenant and the High Sheriff. The former is nearly always a nobleman, or, if not, a large landowner, and is appointed by the Crown. He represents the King in the county, and one of his most important duties is the chairmanship of the committee or association that has control over the Territorial Forces of the county. The sheriff, generally a wealthy man, is chosen annually on the "morrow of St Martin's Day, November 12th."

The County Council conducts the chief business of the county and thus resembles the ancient shire moot. This body was instituted by Act of Parliament in 1888, and in Breconshire is constituted of 15 aldermen and 49 councillors. The councillors are elected triennially by the ratepayers and the aldermen are co-opted by the councillors for a term of six years. Such matters as sanitation, roadways, education, the provision of asylums, water-supply, etc., are under the control of the Council. The County Council holds its meetings at Brecon.

An important committee in the county is the Standing Joint Committee, composed of members of the County Council and of the magistrates of the county, which has the appointment and control of the police. This committee consists of 18 members, and has under its control a body of 34 police of all ranks, who are commanded by a chief constable.

Before the year 1894 there existed a number of local governing bodies known as vestries, local boards, highway boards, etc., but in that year an Act of Parliament was passed which created new bodies to deal with the affairs controlled by these bodies. In the larger and more populous areas they are called District Councils—Urban District or Rural District Councils, as they govern town or rural districts—and the smaller parishes or areas have their Parish Councils or Parish Meetings. These, whether District or Parish Councils, represent the ancient courts of the townships or parishes under the old shire system. In Breconshire there are five Urban District Councils and six Rural District Councils. The parish of Ystradfellte is administered by the Rural District Council of Neath in the Administrative County of Glamorganshire, and the parish of Llanwrthwl by the Rural District Council of Rhayader in Radnorshire. The parishes of Beaufort, Dukestown, Llechryd, and Rassa are in the Administrative County of Monmouthshire. The Borough of Brecon has powers similar to those of the Urban District Councils, and also holds certain privileges under charters granted to it in ancient times.

Breconshire has four Poor Law Unions, each of which has a Board of Guardians, elected by the ratepayers, whose duty it is to manage the workhouses, and appoint the relieving officers and other officials who take care of the poor and aged.

For the purpose of administering justice the county is in the South Wales Circuit, has one court of Quarter Sessions which meets at Brecon, and is divided into ten Petty Sessional Divisions, each having magistrates, or justices of the peace, who attend to try cases and punish petty offences against the law. The Municipal Borough of Brecon has a separate Commission of the Peace but has no separate Court of Quarter Sessions.

By far the greater portion of Breconshire is in the Diocese of St Davids, but three ecclesiastical parishes are in the Diocese of Llandaff. There are altogether 70 ecclesiastical parishes or districts in the county. At one time the ecclesiastical parish was the same as the civil parish, but in modern times changes have been made so that the boundaries do not coincide. There are within the Administrative County 94 civil parishes.

26. Roll of Honour.

Breconshire, sparsely populated, far removed from any great national centre, can hardly be expected to have produced many men who played great parts in the history of the world. Of soldiers, if we except the Norman lords who once held sway in the county and the two Breconshire heroes of Agincourt—Sir David Gam and Sir Roger Vaughan of Tretower—we have none, and the fact that it is an inland county accounts for the lack of famous sailors among its sons.

Among the churchmen of our county, Giraldus Cambrensis—Gerald the Welshman—Archdeacon of Brecon and author of a *Description of Wales* and numerous other works, takes foremost place. This militant cleric, who fought so strenuously for the independence of the Welsh Church, died in 1220, and was buried in St David's Cathedral. Thomas Huet, rector of Cefnllys and Disserth, was the translator of the *Book of Revelations* in Salesbury's edition of the New Testament. He died in 1591. Dr John Price, a divine of the Elizabethan period, founded Jesus College at Oxford.

John Penry, "the Morning Star of the Welsh Reformation," was born at Cefn Brith near Llangammarch about 1554. He was executed in 1593 for his connection with the Martin Marprelate faction. Dr John Jones (1575–1636), the author of several theological treatises and an Exposition of the Bible, was born at Llanfrynach. He was a friend of Archbishop Laud. Thomas Howell (d. 1644) was chaplain to Charles I, and afterwards Canon

Cefn Brith: John Penry's Birthplace

of Westminster and Bishop of Bristol. Thomas Vaughan, of Scethrog (1620–1665), a student of Oriental languages and a poet of some reputation, was incumbent of Llansantffraed. He was twin brother of the "Silurist."

Drych y Prif Oesoedd, a Welsh classic, was the work of Theophilus Evans (1694–1769), incumbent of Llangammarch, in whose churchyard he was buried. Thomas Price, "Carnhuanawc" (1787–1848), the author of *Hanes Cymry* and other works, held the living of Cwmdu

Theophilus Evans

and was an eminent artist, botanist, antiquary and divine. George Granville Bradley, Dean of Westminster and author of *Recollections of Arthur Penrhyn Stanley*, was born at Glasbury in 1822.

Howell Harris and Thomas Coke, two of the best-known of all those connected with the Methodist movement in Wales, both commenced their careers as members of the Church of England. The former, born at Trefecca in 1715, was a great friend of Whitfield, and is regarded as the founder of Welsh Methodism. The latter, born at Brecon in 1747, became a missionary of the Wesleyan Church, and to him was due the establishment of the Foreign Mission Committee of the Wesleyan Church. He died at sea in 1813.

Chief among the literary sons of the county is Henry Vaughan, "the Silurist," who was born at Scethrog in 1620. An M.D. of Oxford, he practised medicine in Brecon and its neighbourhood, but it is upon his literary productions that his fame rests. Of deeply religious temperament he became strongly influenced by George Herbert's poetry and took him for his model in his *Silex Scintillans*. He wrote various works of a religious character, both prose and verse. He died in 1695 and was buried at Llansantffraed. Richard Hall of Brecon wrote *Tales of the Past and other Poems*, and William Churchey, a friend of Southey, was the author of *Poems and Imitations of English Poets*. Dafydd ap Gwilym Buallt, of Builth, a tailor, made a translation of Hervey's *Meditations*, said to be superior in style to the original. Joseph Harris (d. 1775), brother of Howell Harris, was the author of a *Treatise on Optics* and several other mathematical works.

One of the best known and honoured of the sons of our county is Theophilus Jones, its historian. He was the grandson of Theophilus Evans and was born in

1758. His *History of Breconshire* has passed through many editions, and is a comprehensive and valuable account of the county. He died in 1812 and was buried at Llangammarch. The most eminent literary lady of the county is Miss Jane Williams, who was born at Talgarth in 1805 and died in 1885. Her best work was the *Life and Writings of the Rev. Thomas Price, Carnhuanawc*, and she was also the authoress of *Literary Women of the Seventeenth Century* and a *History of Wales*.

The most prominent lawyer of the county is Sir John Price of Brecon, who was a barrister and a member of the Council of the Court of the Marches in the time of Henry VIII. He was the author of a *Defence of British History* and a *Description of Wales*, but he will be remembered by his countrymen as the chief promoter of the Union of Wales with England. Sir John Price died in 1573. Diplomacy is represented by the late Sir H. E. Bartle Frere, who was born at Clydach. He held many important diplomatic posts, chiefly in India, becoming Governor of Bombay in 1862. Sir Bartle Frere was specially sent to South Africa to promote confederation in the terms of Lord Carnarvon's scheme and became the first Governor of Cape Colony. Sir Bartle Frere died in 1884 and was buried in the crypt of St Paul's Cathedral.

The chief representatives of art in the county are three brothers whose name is Thomas. The eldest, John Evan Thomas, F.S.A., was born at Brecon in 1810. He became a sculptor whose works received great praise from the critics. Some of his principal works are a bronze statue of Mr J. H. Vivian, at Swansea; a marble statue of the late King Edward VII when a boy; a bronze statue of the Duke of Wellington which is now in his

Sir Bartle Frere

native town; and a marble statue of the late Prince Consort which is at Tenby. Many fine groups were carved by him, the original model of one, "Science unveiling Ignorance," being in the City Hall, Cardiff. He died in 1875, and his unfinished works were completed

Mrs Siddons

by his brother, William Meredith Thomas. William Meredith Thomas (d. 1877) was a prize medallist of the

Royal Academy of Arts, and was the sculptor of many excellent groups and statues. He was especially noted for his medallion portraiture, which, for a time, was a speciality of his own. The third brother, James Lewis Thomas, was an architect, and being connected with the War Office, was the designer of the Royal Victoria Hospital, Netley, and other public buildings.

The stage is ably represented in the persons of the gifted tragic actress, Mrs Sarah Siddons, whose family name was Kemble, and her youngest brother, Charles Kemble, father of Fanny Kemble, who was an actor of great repute though not of the rank of his eldest brother, John Philip. These talented players were born at the Shoulder of Mutton Inn, now the Siddons Vaults, in High Street, Brecon, Mrs Siddons in July 1755, and her brother in 1775. Mrs Siddons also wrote some verses and published an abridged edition of *Paradise Lost*. Another actor that may well be mentioned here was T. L. Llewelyn Pritchard, but he is better remembered by his countrymen as the author of *Heroines of Welsh History*, and of the novel *Twm Shon Catti*. He was born at Trallwng at the beginning of the nineteenth century, and is supposed to have died in poverty at Swansea.

27. THE CHIEF TOWNS AND VILLAGES OF BRECONSHIRE.

The figures in brackets after each name give the population in 1901, except in the cases where the figures for 1911 are available when they are distinguished by an asterisk. The figures at the end of each section are references to the pages in the text.

Abercrave. See under Ystradgynlais.

Aberyscir (117) is a parish and village on the west bank of the Yscir at its confluence with the Usk. The church was restored in 1860 and 1884. On the river bank near the church are the remains of the old Roman station of *Bannium*. (pp. 18, 97, 109, 110, 129.)

Alltmawr (38) is a parish on the right bank of the Wye on the road from Builth to Hay some four miles south-east of Builth. Its parish church is one of the smallest in the Principality.

Battle (113) is a parish and village named after the great Abbey of Battle in Sussex. (pp. 97, 102, 109.)

Brecon (*5908) is a municipal borough and market town and the chief town of the county. It was formerly the *caput* of the lordship of Brecon and is situated on the spot where the Honddu pours its waters into the Usk. This gives it its Welsh name of Aberhonddu. The Shire Hall is a handsome building in the Doric style and there is also a Guild Hall. The

Christ College, Brecon

town contains the County Prison and is the depôt of the 1st and 2nd Battalions of the South Wales Borderers. Fragments of the ancient castle are to be seen in the grounds of the Castle Hotel. The "Castle Inn" is a civil parish in the county and comprises the precincts of Brecon Castle and was formerly extra-parochial. There are only two houses in the parish, one of which, the Castle Hotel, gives it its name. The town has Christ College, a public school, which however is not included in the borough boundaries; a Memorial College for Congregational ministers, and several handsome churches. A bronze statue of the great Duke of Wellington by J. E. Thomas, F.S.A., erected in 1854, stands facing the Bulwark. There are numerous remains of antiquity in the neighbourhood. The industries of the town include brewing, tanning, and the manufacture of flannel and tweed. (pp. 19, 46, 47, 66, 81, 86, 88, 91, 97, 102, 106, 109, 110, 113, 133, 135, 136, 137, 141, 146, 147, 150.)

Bronllys (304) is a parish and village on the road from Brecon to Hereford, eight miles from the county town. The village is supplied with water from works the property of the ratepayers. Malting is carried on in the parish. Near the church is the site of a British camp and there are the remains of Bronllys Castle. (pp. 92, 106, 115.)

Brynmawr (*civil par.=7592, eccles. par.=6543) is a town and ecclesiastical parish on the border of the county. It derives its water from a reservoir on Cryn Mountain which is the property of the District Council. The town was once an important seat of the iron and coal industries, but the coal is exhausted and the iron ore is not suitable for the present mode of manufacturing steel. (pp. 11, 68, 77, 107.)

Builth (*1710) is an inland watering-place and a market town on the banks of the river Wye. Owing to its agreeable climate and celebrated mineral springs it is a favourite resort for invalids and others. Surrounded by some of the most magnificent mountain scenery in South Wales, its situation is an ideal one. A six-arched stone bridge crosses the river into Radnorshire. The town consists chiefly of one main street. It is well paved and is supplied with water from two springs, which provide water for two storage reservoirs holding 122,000 gallons. The river is navigable for boats, and good trout fishing forms an additional attraction to the town. The site of an ancient castle of eleventh century date stands on the east side of the town. Near the town is Cwm Llewelyn, a wooded ravine where Prince Llewelyn fell in 1282. Aberedw rocks and caves are also in the vicinity. Nant yr Arian, "Money Brook," is to the south-west of the town and is so called from the days of the plague. Here the country people brought their produce for sale, the money in payment therefor being thrown into the brook to prevent any possible contagion. Builth has also experienced several destructive fires, one in 1691 completely destroying the town. The Builth Philharmonic Society, formed by the amalgamation of the old Builth Harmonic Society and another choir, is held in high repute. Both choirs won the chief choral prize at the National Eisteddfod of Wales. (pp. 8, 12, 23, 39, 41, 50, 67, 70, 81, 82, 83, 85, 89, 90, 92, 99, 109, 115, 135, 137, 146.)

Bwlch. See under Llanfihangel Cwmdu. (pp. 44, 50, 113, 133.)

Callwen. See under Glyntawe.

Cantref (185) is a parish lying two and a half miles south-east of Brecon. The Brecon Beacons, the highest mountains in South Wales, lie about the centre of the parish; Pen y Fan, the highest peak, reaching a height of 2907 feet. (p. 76.)

Cefn Coed y Cymmer is a hamlet and large village in the parish of Vaynor two miles north-west of Merthyr Tydfil. The river Taff runs near the village and over it is a viaduct of

16 arches which cost £25,000 to build. The span of each arch is 40 feet and the height from river to parapet is 120 feet.

Cray, or **Llanulid** (709) is a large and scattered parish once part of the parish of Devynock. It is 12 miles from Brecon. In the Cray Valley is a large reservoir for the supply of water to the town of Swansea. It was constructed at a cost of £800,000, and comprises a concrete dam across the valley,

Vaynor Viaduct

and a tunnel three miles in length which runs underneath a mountain, terminating at Nantyrwydd, whence water is conveyed in pipes to Swansea. The dam is over a quarter of a mile long and keeps back over 1000 million gallons of water, which cover an area of 150 acres. This parish forms the greatest portion of the site of the once great Forest of Brecon. (p. 76.)

Crickadarn (295) is a strictly agricultural parish eight miles south-west of Builth.

Crickhowell (1150) is a market town and parish lying in a picturesque valley on the banks of the river Usk, over which is a bridge of 13 arches. The river is wide here and provides good fishing. There are many remains of antiquity in the neighbourhood and the curfew is rung at Crickhowell from Michaelmas to Candlemas. There are several public buildings, chief of which are the Town Hall and the Clarence Hall. The first stone of the latter was laid by the late Duke of Clarence in 1891. There are

Devynock

the remains of a castle, and at Llangattock (924) is a remarkable cavern called "Eglwys Vaen" or the Church Cavern. There are also several remarkable monoliths in the vicinity and the remains of a cromlech. (pp. 3, 20, 46, 65, 66, 91, 93, 107, 109, 110, 117, 127, 133, 135.)

Cwmdu. See Llanfihangel Cwmdu. (pp. 20, 37, 145.)

Devynock (1027) is an ecclesiastical parish comprising several townships. The village of Devynock stands in the township of Maescar and is nine miles west of Brecon. There is a small woollen-factory here and a corn-mill and saw-mills. Some business is done in corn, coal, flour, etc. at Senni Bridge. There are the remains of several camps in the neighbourhood. (pp. 66, 86, 90, 92, 101, 109, 110, 121, 137.)

Eglwys Oen Duw (468) a village in a wild and mountainous district on the main road from Builth to Llanwrtyd. The people are mainly agriculturists.

Glasbury is a parish and village situated on the banks of the Wye in Radnorshire and Breconshire. The ecclesiastical parish of All Saints is on the Radnorshire side of the river and the ecclesiastical parish of Glasbury is in Breconshire. The former has a population of 460 and the latter of 720. Tregoyd and Felindre are hamlets in the parish of Glasbury with a total population of 490. (pp. 39, 91, 109, 145.)

Glyn. See under Libanus.

Glyntawe, or **Callwen** is an ecclesiastical parish situated around the source of the Tawe. At **Penwyllt**, a village in the parish, are fire-brick and lime works, and limestone quarries. In the neighbourhood are remains of antiquities and the limestone rocks contain caves of varying interest. The population numbers 147.

Hay (*1603) is a market town on the borders of Breconshire, Radnorshire, and Herefordshire. The town occupies rising ground on the south bank of the Wye, which is here crossed by an iron bridge connecting Radnorshire and Breconshire. The town has several main streets and some good houses and shops, but the trade is purely agricultural. The Town Hall serves as a cheese-market. There are remains of a castle in the town, a portion of the site being occupied by the Glanusk dower house. The town owes its origin to the Normans and the name is an Anglicised form of La Haie (the enclosure) and is still referred to as "The Hay." (pp. 8, 11, 14, 22, 37, 47, 50, 66, 69, 82, 85, 110, 119, 135, 137.)

Libanus, Llanelltyd, or **Glyn** (206), a hamlet and ecclesiastical parish which comprises the township of Glyn, is on the river Tarrel four miles south-west of Brecon. A Roman road ran through the parish and the reputed site of St Illtyd's grave is on the side of Illtyd Mountain.

Llanafan Fawr (457) is a parish six miles north-west from Builth. In the churchyard of St Afan is an altar tomb with the inscription, "*Hic Jacet Sanctus Avanus Episcopus.*" In a field on a farm (Dolfelin) stands an ancient monolith believed to mark the site of the grave of a bishop slain by the Danes, and near this runs a stream known as "Nant yr Esgob." **Llanafan Fechan** is a parish lying south of the former. It has a population of 129.

Llanbedr (204) is a small parish and village near the confluence of the rivers Grwyne Fawr and Grwyne Fechan, some two miles from Crickhowell. Since 1865 the village has been almost entirely rebuilt. (p. 20.)

Llanddetty (458) is a parish and scattered village six miles from Crickhowell. The church of St Detta, an ancient building in the Early English style, was restored in 1874. The village of Talybont is partly in this parish. (p. 99.)

Llanddew (183) is a parish and village on the river Honddu, one and three quarter miles from Brecon. The manor is attached to the bishopric of St Davids, and there was once a castellated mansion here of which some traces remain in the vicarage grounds. It was the Breconshire home of the famous cleric and writer, Giraldus Cambrensis.

Llanddewi'r Cwm (426) is an agricultural parish and village two miles south of Brecon.

Llandefalle (446) is a parish seven miles north-east of the county town. Pen-heol-Einion in this parish was the scene of a tenth century battle. There are the remains of an ancient

Llanelieu Church

encampment at Pwll Cwrw. The church has the remains of an ancient rood screen of great beauty. (p. 109.)

Llanelieu (53) is a small parish two miles east of Talgarth. The church contains an ancient double screen of the fourteenth century, the eastern and western faces being similar, and the interval between them roofed over so as to form a rood-loft. The socket of the ancient rood remains in the beam, a painted cross now taking its place. (p. 109.)

Llanelltyd. See under Libanus.

Llanelly (3076) is a parish three miles from Brynmawr and consists of the villages of Blackrock, Clydach, Darranfelen, Gilwern, and Maesgwrtha. Llanelly was once a busy mining district but there are now no works of any kind except limestone quarries and kilns at Clydach. The limestone here yields 97 and 98 per cent. of lime. Sir Bartle Frere, first Governor of Cape Colony, was born at Clydach. (pp. 66, 107, 147.)

Llanfihangel Cwmdu (826) generally known as Cwmdu and originally called Ystradyw, is a large and scattered parish on the river Rhiangoll about four and a half miles north-west from Crickhowell. **Bwlch** is a small village situated in this parish. At Gaer, or Coed y Gaer, in the parish are the remains of an ancient encampment, and on the Brianog Mountains are traces of British hut dwellings. (pp. 93, 97, 105, 109, 133.)

Llanfihangel Fechan or **Llanfechan** (126) is a parish five miles north from Brecon. The parish church is plain but has an apsidal chancel. Castell Madoc is in this parish and there are traces of a British camp called Twyn y Gaer.

Llanfrynach (462) is a parish and village on the river Usk, three and a half miles south-east from Brecon. A Roman bath and coins were discovered here, and in 1898 a kistvaen, or stone sepulchral chest, containing human remains, was discovered in a field called "Cae Gwyn" on the farm of Ty'n Llwyn. (pp. 93, 98, 144.)

Llangammarch (688) is a small town situated amongst a charming tract of hilly and well-watered country through which flows the celebrated trout stream the Irfon. The parish comprises the townships of **Penbuallt** (433) and **Treflis** (481) and lies about four miles east of Llanwrtyd and eight from Builth. At **Llangammarch Wells** are the only barium springs in the country and the wells extend, with their attendant houses, along the valley of the Irfon. Llangammarch Wells is rapidly growing into favour as an inland watering-place. (pp. 23, 52, 70, 99, 137, 143, 145, 147.)

Llanganten (205) is a parish three miles west from Builth. In this parish at a spot called "Cefn y Bedd," Prince Llewelyn ap Gruffydd, the last native Prince of Wales, is said to have been slain. (p. 85.)

Llangattock. See under Crickhowell. (pp. 66, 79.)

Llangenau, or **Llangenny** (347), is a parish two miles south-east from Crickhowell. The village of Llangrwyne is in this parish, which is interesting as containing two parish churches. There is an iron bridge here crossing the Usk, and the Sugar Loaf mountain is in the parish. Here are paper and cardboard mills, and in the park of Court y Gollen (destroyed by fire in 1909) is a menhir. (p. 94.)

Llangorse (327) is a parish and village two miles west of Talyllyn Junction and four and a half miles south of Talgarth. Llangorse Lake lies in this parish. It is three miles long, one mile across, and five miles in circumference. It lies in a low and marshy tract in the valley of the Llyfni and is a favourite resort of anglers. (p. 99.)

Llangynidr (479) is a parish and scattered village four miles from Crickhowell. The Usk is here crossed by an ancient bridge of six arches. Limeburning is carried on in the parish. The church is an ancient edifice in the Early English style and was restored in 1872 and 1895. (p. 77.)

Llanhamlach (249) is a parish on the river Usk. The church of St Illtyd and St Peter was, with the exception of the

Henry Vaughan's Tomb

tower, rebuilt in 1802. The architecture is plain in the Norman style. On Mannest, a hill in the parish, are the remains of a dolmen or cromlech known as Ty Illtyd, the stones of which once had an inscription which has disappeared, and near is Maen Illtyd, the remains of an ancient stone circle. (pp. 98, 110.)

Llanigon (Civil par. = 291, eccles. par. = 336) is a parish and village two miles south-west from Hay. About a quarter of a mile distant from Glynfach, a hamlet in the parish, is the newly-erected Benedictine Monastery of the late Father Ignatius.

Llanlleonfel (94) is a parish on the river Dulais seven miles from Builth. The Garth brickworks are in the parish, and at Garth there is also a recently-discovered magnesium spring. There are also some Roman remains in the parish. (p. 97.)

Llansantffraed-juxta-Usk (191) is a parish six miles south-east from Brecon. **Scethrog,** the birthplace of Henry Vaughan, the "Silurist," is the south-western portion of the parish and derives its name from Brochwell Yscythrog, one of the Princes of Powys, and still includes a group of houses called by this name. About half a mile from the cottages is a pillar stone, about three feet in height, with a much mutilated inscription, of which only the word "Victorini" is now legible. Tradition states that it commemorates the burial place of a son of Victorinus who was slain in a neighbouring dell. The Roman road from Caerleon to Y Gaer (Brecon) ran through the parish. At Buckland, on a hill known as Buckland Hill, the late Colonel Gwynne-Holford, an old Waterloo veteran, had trees planted to represent squares of infantry and troops of cavalry at Waterloo. (pp. 19, 90, 109, 144, 146.)

Llanspyddid (civil par. = 160, eccles. par. = 249) is a parish two miles south-west from Brecon. The church is of thirteenth century date and has a finely carved sounding-board over the pulpit, and the east window is a memorial to the Morgan family. In the churchyard are 13 fine yew trees supposed to date from the fifth century, and there is also an ancient stone said to mark the grave of Awlach, one of the princes of Brycheiniog. (p. 101.)

Llanulid. See Cray.

Llanwrthwl (1785) is a parish in an agricultural district on the northern extremity of the county, close to the Radnorshire boundary. The parish church contains a twelfth century font, and in the churchyard is a fine menhir. On Rhos Saith Maen (the meadow of the seven stones) are some stones irregularly placed, sometimes designated a stone circle, but their origin and purpose cannot be clearly determined. (p. 142.)

Llanwrtyd, or **Llanwrtyd Wells** (854) is a large village and parish 11 miles west from Builth. It stands on the river Irfon, a celebrated trout stream, in the midst of the wild and picturesque scenery. The chief interest of the place lies in its celebrated sulphur and chalybeate springs. There are three of the former, which are said to be the strongest sulphur springs in the Principality. Many hotels and lodging-houses cater for the visitors, and there are pleasure grounds and a small lake fed by the Irfon. (pp. 23, 36, 37, 66, 70, 90, 92, 137.)

Llywell (654) is an ecclesiastical parish in an agricultural district on the road from Brecon to Llandovery. The church is an ancient building of red sandstone and is dedicated to St Padarn, St Teilo, and St David, hence the alternative name Llantrisant. **Trecastle** (380) is a hamlet in this parish, as is **Ysclydach** or **Rhyd-y-briw** (202). (p. 101.)

Merthyr Cynnog (590) is an agricultural parish nine miles north by west from Brecon. A little flour-milling by water power is carried on. As the name implies, it was the scene of the martyrdom of Cynnog, a Celtic Saint. (p. 109.)

Newbridge on the Wye (Rad.=404, Brec.=183) is a parish partly in Radnorshire and partly in Breconshire, with some trade in lime and coal.

Patricio, Partishow, or **Partrishow** (43) is a small parish on the Grwyne Fawr six miles north-east of Crickhowell.

The church of St Ishow is a Perpendicular structure with earlier remains. The chief feature of the church is a beautiful rood screen and loft in the florid Gothic style. It is one of the finest examples in the country and is splendidly carved. It is supposed to be the work of an Italian monk and to have been presented to the church by one of the Herbert family in the fourteenth or fifteenth century. Below the rood-loft, two stone pre-Reforma-

Rood Screen, Partrishow Church

tion altars retain their original positions and one table slab still exhibits several consecration crosses. The font here is an interesting example and the church also possesses a black letter Welsh Bible dated 1620. Near the church is a Holy Well. (pp. 107, 108, 109.)

Penderyn (civil par.=1346, eccles. par.=607) is an extensive parish on the border of the county, seven miles west of

Merthyr Tydfil. Here are limestone, sandstone, and silica quarries. (pp. 66, 77, 92.)

Penwyllt. See under Glyntawe.

Pont Neath Vaughan (Nedd Fechan) is a village partly in Glamorganshire and partly in Breconshire. It stands in the vicinity of magnificent scenery and forms the centre whence tourists visit the many beautiful waterfalls and other interesting natural features of the district. Quarrying forms the chief local industry with some coal mining. There are silica and limestone quarries and on the Mellte are gunpowder works. In former days numbers of Welsh hats were made here. (pp. 11, 15.)

Scethrog. See Llansantffraed. (pp. 97, 110, 121, 144, 146.)

Talgarth (1466) is a parish and market town seven and a half miles south-west from Hay. The Vale of Talgarth is one of the most important and richest agricultural districts in the county and the town has an important and well-attended horsefair. It is supplied with water from a reservoir on the Black Mountains. The Market Hall, erected in 1877, is in the centre of the town. The upper part is used as an Assembly Room and the lower as a provision market, etc. The town is a great centre for markets and fairs, and depends entirely for its trade upon the agricultural district around. At Great Porthmael, once a fortified residence, now a farmhouse, Henry VII is said to have rested on his way to the battle of Bosworth Field. Dinas Castle is near the town, and there is an ancient stone circle at Ffostill. Trefecca College stands about a mile from the town. The college was originally the home of a community of persons who settled here under the governance and teaching of Howell Harris. It was later a college for Calvinistic Methodists and was opened for this purpose in 1842. The college library has a valuable collection of Welsh books. In the grounds is the

Howell Harris Memorial Chapel, which was opened in 1873. The county asylum is at Talgarth. (pp. 23, 39, 67, 89, 91, 107, 109, 121, 137, 139, 147.)

Traianglas (434) is a hamlet and ecclesiastical parish near the confluence of the Hydfer with the Usk. Both streams are favourite waters with anglers and contain trout and salmon. On Trecastle mountain in this parish is a Roman camp, and coins and other Roman remains have been found from time to time. A large lake is situated on the Fan Mountain.

Trecastle (380) is a hamlet in the ecclesiastical parish of Llywel some four or five miles from Devynock. At the New Factory some wool-dressing is carried on. **Ysclydach** (202) is another hamlet in the same parish. (pp. 17, 92, 135.)

Tretower (348) is an ecclesiastical parish three miles north from Crickhowell. There are the remains of a medieval castle here, and immediately adjoining it is a manor house of the time of Edward III. (pp. 20, 92, 95, 97, 122, 126, 133, 143.)

Vaynor (3420) is a parish four miles north of Merthyr Tydfil. Part of the parish is in the parliamentary borough of Merthyr Tydfil. **Cwmtaff** and **Dyffryn** are hamlets and the populous village of **Cefn Coed y Cymmer** is in the parish. The chief industries apart from agriculture are quarrying and lime-burning. Near Vaynor church is an exceptionally fine tumulus and there are others in the vicinity. (pp. 92, 95, 107.)

Ysclydach. See under Llywell.

Ystradfellte (549) is a purely pastoral parish on the road from Brecon to Neath. The village of Pont Nedd Fechan is in this parish. Reservoirs belonging to the Borough of Neath are now in process of construction in this district. (pp. 27, 92, 136, 142.)

Ystradgynlais (Higher=820, Lower=4965) is a parish in the south-western part of the county. It contains the villages of Ystradgynlais, Abercrave, Cwmtwrch, and Gurnos, and its industries include anthracite coal mining, tin-plate manufacturing, cement-works and brick-making. Scwd yr Hen Rhyd waterfall, one of the most attractive in Wales, is in the parish. The Swansea Canal has its northern terminus here. There are traces of a Roman road and the remains of a Roman camp to be seen. (pp. 16, 66, 68, 129.)

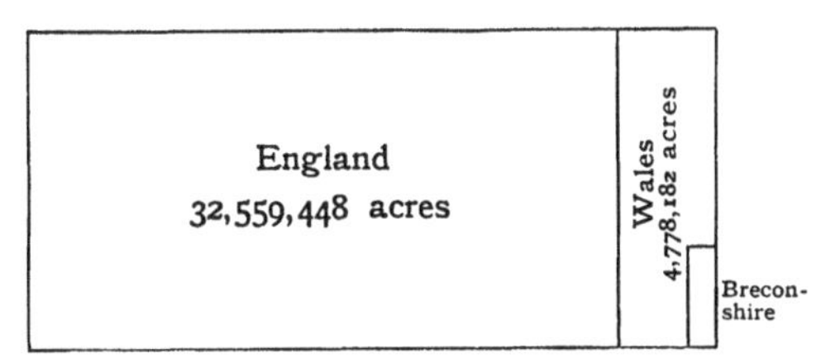

Fig. 1. Area of Breconshire (469,281 acres) compared with that of England and Wales

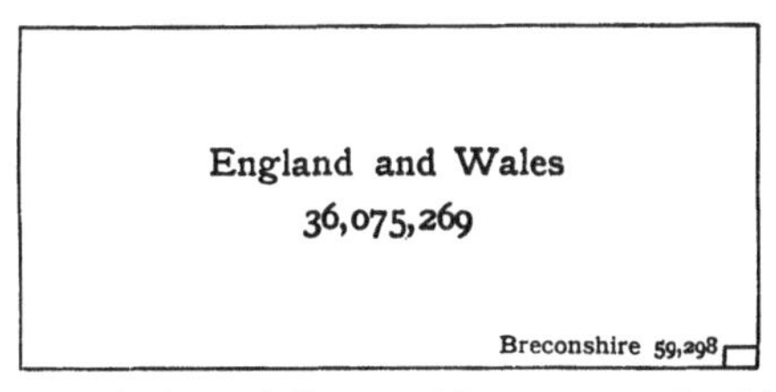

Fig. 2. Population of Breconshire compared with that of England and Wales in 1911

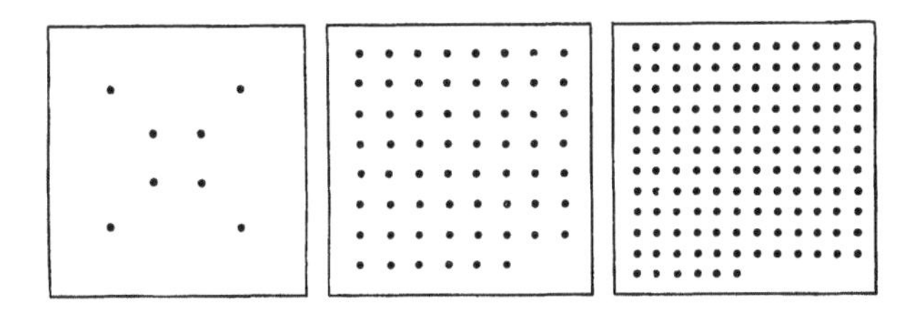

Fig. 3. Comparative density of population per square mile in Breconshire, England and Wales, and Glamorganshire in 1911

(Each dot represents 10 persons)

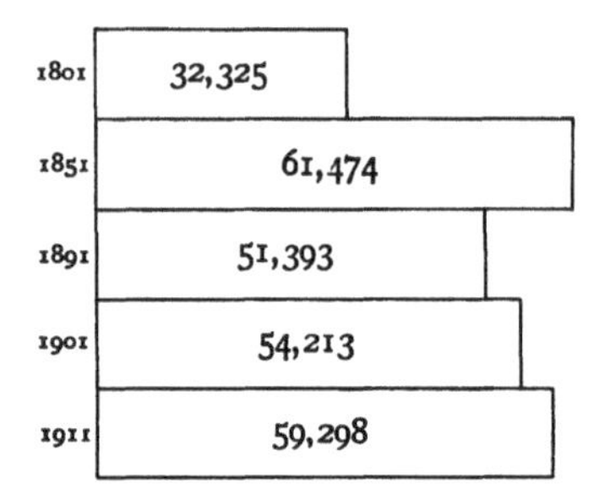

Fig. 4. Variation in population of Breconshire from 1801 to 1911

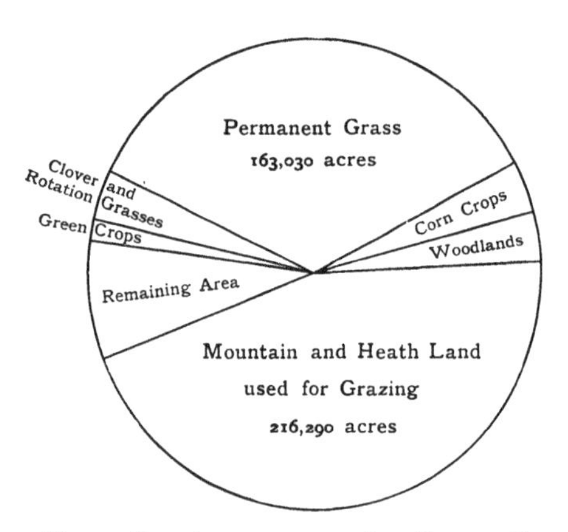

Fig. 5. Proportionate acreage under Crops, Grass, etc. in Breconshire in 1909

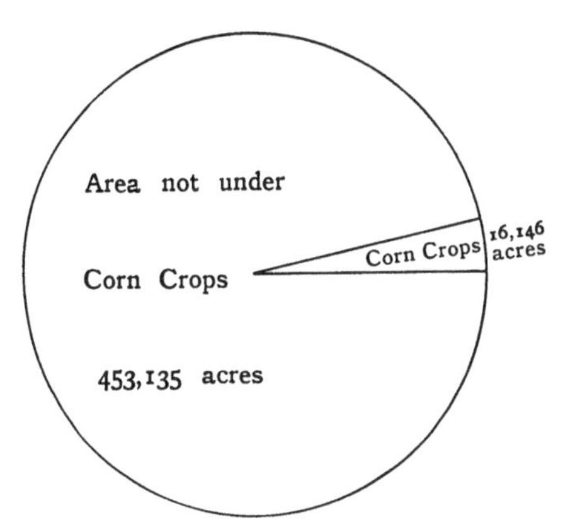

Fig. 6. Area under Corn Crops in Breconshire in 1909

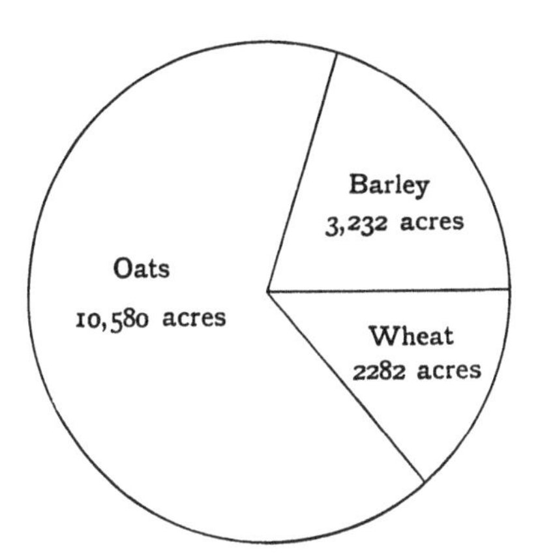

Fig. 7. Proportionate areas of chief Cereals in Breconshire in 1909

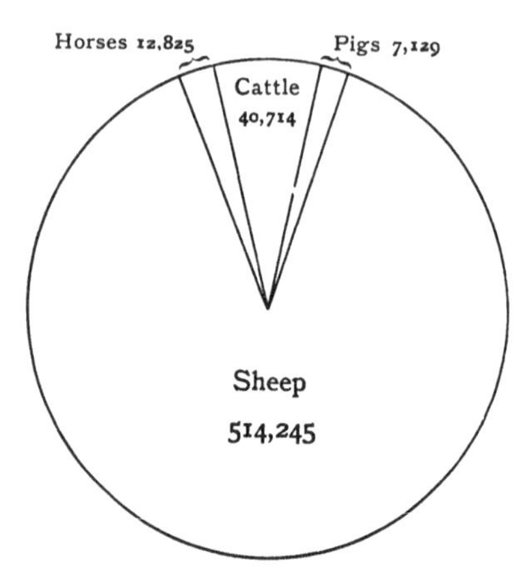

Fig. 8. Proportionate numbers of Live Stock in Breconshire in 1909

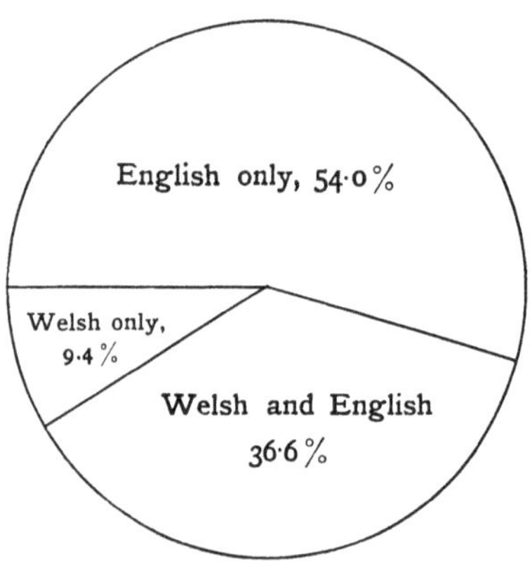

Fig. 9. Diagram showing percentage of English-speaking, Welsh-speaking, and Bilingual Persons in Breconshire in 1901

www.ingramcontent.com/pod-product-compliance
Ingram Content Group UK Ltd.
Pitfield, Milton Keynes, MK11 3LW, UK
UKHW042143280225
455719UK00001B/74